COUNTDOW

ENGLISH

Patrick Scott

Chair of the National Association
for the Teaching of English

M
MACMILLAN
EDUCATION

First published 1986

Published by
MACMILLAN EDUCATION LTD
Houndmills, Basingstoke, Hampshire RG21 2XS
and London
Companies and representatives
throughout the world

Designed and illustrated by
Plum Design
Southampton

Typeset by TecSet Ltd, Sutton, Surrey

Printed in Great Britain by
Cox & Wyman Ltd
Reading

British Library Cataloguing in Publication Data
Scott, Patrick, 1949–
English. — (Countdown to GCSE)
1. English language — Grammar —
1950–
I. Title   II. Series
428   PE1112
ISBN 0-333-41345-8

ACKNOWLEDGEMENTS

The author and publishers wish to acknowledge the following sources:

Department of Health and Social Security for extract from leaflet SBI 'Cash Help'; Durham County Council for extract from timetable, May 1985, by permission of the County Planning Officer; Glen Hughes' Yorkshire for extract from *Millstone Grit Revisited* published by Chatto and Windus; Grafton Books for 'Ninetieth Birthday' taken from *Tares* by R. S. Thomas; HMSO for extracts from the National Criteria, by permission of the Controller of Her Majesty's Stationery Office; Hamish Hamilton and Penguin Books for extract from *I'm the King of the Castle* by Susan Hill (1970); Hodder & Stoughton Ltd for extract from *Frugal Food* by Delia Smith; Michael Joseph Ltd for extract 'Smoking' from *Sunday Times Book of Body Maintenance*, 1978, ed. Oliver Gillie, Celia Haddon and Derrik Mercer, and for extracts from *Kes* by Barry Hines; Laboratoires Roc UK Ltd for permission to reproduce an advertisement; Longman Publishers for extract from *Crime and Punishment* by Robert Roshier (1976); *The Observer* for extracts 'Cooling Towers', March 1985 and 'Last in the Class' March 1985; Penguin Books Ltd for extract and illustration from *Miss Brick the Builder's Baby* by Allan Ahlberg and Colin McNaughton (Puffin Books and Kestrel Books 1981), page 2 text © Allan Ahlberg 1981, illustration © Colin McNaughton 1981; Laurence Pollinger Ltd for extracts from H. E. Bates *The Triple Echo*, published by Michael Joseph Ltd, by permission of Laurence Pollinger Ltd and the Estate of H. E. Bates; also
London and East Anglian Group; Midland Examining Group; Northern Examining Association; Southern Examining Group.

The publishers have made every effort to trace the copyright holders, but if they have inadvertently overlooked any, they will be pleased to make the necessary arrangements at the first opportunity.

CONTENTS

Countdown to GCSE: English

SECTION I

Introduction — why GCSE?

If you are being entered for the GCSE and fear that you are being used as a guinea-pig for a new set of exams, you might be reassured to know that the first discussions about merging 'O' level and CSE took place before you were even born. The new exams aren't happening because somebody woke up one morning with a bright idea. They have been in preparation for a long time, and the syllabuses that are being introduced rely very heavily on methods of teaching and examining that have been thoroughly tested in schools all over the country.

That's fair enough, you might say, but why bother? The system worked well enough, so why change it? If that is how you feel, then you might be surprised to learn that, in English at least, 'O' level and CSE came in for some very fierce criticism indeed over the last twenty years from people who ought to know what they are talking about.

A government-sponsored report on *The Examining of English Language at 'O' level*, which was published in 1964, had this to say:

> We have considered most seriously whether we should advise the cessation of these exams for educational reasons . . . we have come very near that conclusion.

Ten years later the verdict was much the same, though the Bullock Report put it a bit less bluntly:

> English requires a wider and more flexible range of assessment . . . we believe that rigid syllabuses are not the best means of achieving this.

Even Her Majesty's Inspectors, who spend a lot of time observing what takes place in schools, remained unconvinced:

> The schools and boards responsible for conducting examinations might consider jointly whether exam requirements could be framed so as to encourage more effective learning and use of language.

Perhaps the question that you should be asking is not: 'Why are things being changed now?' but: 'Why weren't they changed a long time ago?'

1

These criticisms weren't ignored completely. New syllabuses were devised, and new methods of assessment tried out. But because some exam boards moved more quickly than others, it soon became apparent that 'English' didn't always mean the same thing in different parts of the country. GCSE is, in part, an attempt to remedy that situation by bringing everybody up to date. What it means in practice is that you will have to get used to new kinds of work in English lessons.

Countdown to GCSE: English is designed to help you do this, by explaining how the exams will operate and by leading you through some of the kinds of assignments and questions that you are likely to be set. If you are going to understand all the terms that are used in the book, however, there are one or two things that you need to know first about how the GCSE is being introduced.

In many parts of the country, *pilot schemes*, called '16+' or 'Joint GCE/CSE' exams, have been running for ten years or more. Since many of the new GCSE syllabuses are based on this experience, you'll find that some of the examples used in the book are taken from those pilot schemes, which were administered by 'O' level and CSE boards, working in partnership. For the GCSE, all the exam boards, and there were 22 of them, have been grouped together into five larger regional examining associations — four in England and one in Wales. They are often referred to by their initials:

NEA – Northern Examining Association;
MEG – Midland Examining Group;
SEG – Southern Examining Group;
LEAG – London/East Anglia Group;
WJEC – Welsh Joint Education Committee.

Although they are all producing their own syllabuses, they don't have complete freedom to do what they want. They have to stick to a set of rules which were drawn up by all the boards working together and approved by the Government. These rules are called the *National Criteria* and you'll find them being referred to quite often in the book. The point about them is that they lay down minimum requirements for each syllabus, and they apply to all candidates, whatever syllabus they may be taking, or whatever part of the country they are living in.

Because there's still room within these rules for there to be considerable differences between one examining group and another, however, *Countdown to GCSE: English* doesn't concentrate on any particular syllabus. It has been designed as a general introduction to the kinds of changes that are taking place, and it is organised in sections: 'Writing', 'Reading', 'Oral Communication' and 'Literature'. Within each section you will find that the examples provided cover all the main types of questions or tasks that will be found in GCSE. Not all of them will apply to every candidate, so the best way to use the book is to pick out the bits that you need to read in greater detail. You will also find that the book aims to take you through some of the reasons *why* the GCSE has turned out the way it has. This approach raises some obvious problems, the main one being that teachers and examiners use a specialist language to talk about what they do that is often difficult to understand unless you are familiar with it. Indeed a great many teachers avoid discussion of this kind for the excellent reason that it might only be confusing. However, it's usually true that if you understand why you are doing something, you are likely to do it better, and English exams are no exception to this. Don't ignore the sections, then, that introduce the aims and objectives of each part of the exam. They could make all the difference, and exercises have been provided to help you make sense of them.

SECTION 2
English: Aims and objectives

This section is called 'English' rather than 'English Language' because the National Criteria say that:

The subject ENGLISH is to be regarded as a single unified course leading to an assessment in ENGLISH. It may also lead to a separate assessment in ENGLISH LITERATURE.

What that means is that there will no longer be any more exams in English Language. After all, the argument goes, why should literature be excluded, since it is one of the most common of the many ways in which the language is used. English Literature as a separate examination is dealt with later.

These, then, are the 'Assessment Objectives' for *English*:

2.1 The Assessment Objectives in a syllabus with the title *English* must provide opportunities for candidates to demonstrate their ability to:

2.1.1 understand and convey information;

2.1.2 understand, order and present facts, ideas and opinions;

2.1.3 evaluate information in reading material and in other media, and select what is relevant to *specific purposes;*

2.1.4 articulate experience and express what is felt and what is imagined;

2.1.5 recognise implicit meaning and attitudes;

2.1.6 show *a sense of audience* and an awareness of style in both formal and informal situations;

2.1.7 exercise control of appropriate grammatical structures, conventions of paragraphing, sentence structure, punctuation and spelling in their writing;

2.1.8 communicate effectively and appropriately *in spoken English*.

Stated baldly like that, they sound a bit formidable. Mention exams to most people, and they think of the actual question papers that they had to answer. It's difficult to make the connections between those papers and a list of numbered objectives. Connections there are, though, and the words that are used in the National Criteria are all highly significant for English teachers.

One of the best ways of making sense of a difficult piece of writing is to take a marker pen to it and highlight the important words. By treating the objectives in this way, it's possible to spot what changes they are likely to promote.

Objective **2.1.3** contains the phrase 'specific purposes', and, later on, in objective **2.1.6**, there is a reference to showing 'a sense of audience'. Those two ideas are closely connected and, together, they sum up one of the biggest changes that has taken place recently in people's thinking about how to write. Here's part of a question from an 'O' level paper set in 1961:

Choose one of the following subjects for composition. About one hour should be spent on this question.

. . . (d) Science in the service of agriculture or building or aviation.

If you thought that was a bit too difficult, you could always go for one of the other options — **(a)** 'The importance of the wheel' — for instance. One look would probably be enough to send you scurrying back to your first choice.

That these questions seem so difficult is not because standards at 'O' level have fallen during the intervening years. It's because it's very hard to make much sense of them in terms of what we now understand about writing. That's why the two words 'purposes' and 'audience' are so important. The way the questions were set in that 1961 'O' level exam didn't provide many clues about *why* you were being asked to write about science or the importance of the wheel, or *who* might read it. And yet whenever, as real writers with something to communicate, we put pen to paper, it's virtually impossible to say anything until we know who it is intended for. GCSE syllabuses face up to that, and provide more realistic conditions both for writing and for reading.

However, what is perhaps the biggest change can be found in objective **2.1.8**, which makes plain that 'spoken English' must be included in GCSE syllabuses. With very few exceptions, the GCE boards confined their 'O' level English Language exams to essay writing and 'comprehension'. This wasn't because they were hostile to spoken English, but because they felt it was difficult, if not impossible, to test performance accurately. The reason why it has been included in the GCSE is not because those doubts have disappeared, but because it is now generally recognised that what gets taught is what gets examined and 'oral communication' is too important to be left to chance or the enthusiasm of individual teachers. The reason why talking is now considered to be as important as writing and reading is straightforward enough — it's because most of the language that we use in adult life is spoken language. That remains true even if you look solely at the world of work. When the Education Authority in Coventry investigated the 'oral and written communication which takes place in industry and commerce' this is what they found:

> Of the three main modes of language used, oral communication was by far the most important.

and

> ... less attention is paid publicly to oral communication, though it appears to be of greater importance to the success and wellbeing of school leavers than the ability to spell and write.

We have concentrated on two features of the objectives, not because the others can be dismissed but because these two emphasise some of the main differences between 'O' level and GCSE. Before moving on, however, it's worth drawing attention to a brief comment that is provided at the end of the list of objectives:

The skills listed above are clearly inter-related and interdependent and, whilst all must be assessed, it is not envisaged that each skill need be separately tested.

In other words, don't expect to find that each English lesson you attend will isolate a particular objective and deal with that — spelling, for example, or oral communication. The best lessons will encourage you to move naturally from one kind of language use to another — from reading to talking, for example — so that you have a chance to try out your skills in a way that matches what happens in the world outside school.

SECTION 3

Writing

INTRODUCTION

One big change that the GCSE will bring is that part, if not all, of your marks will be awarded not for the work that you do in a final exam, but for the work you do during the course. The reason for this goes back to the ideas about 'purposes' and 'audiences' for writing that were touched on in the section on 'aims and objectives'. If the examiner wants to discover whether you can change the way in which you write to suit different readers, then one essay, done under exam conditions, isn't going to be much help. The solution is a 'folder' or 'folio' of coursework, containing a number of pieces of writing of different kinds. It's likely to be very important to make sure that each essay is obviously different from all the others, and so you will need to develop some sense of what is meant by one phrase used in the National Criteria: 'a variety of styles of writing'.

Here is a selection of four different kinds of writing. Can you see the differences between them?

Safe, gentle make-up from RoC.

RoC beauty products from France were created for discerning women.

If you wish to look beautiful, and keep your skin healthy as well, this superb unperfumed range for eyes, lips and nails is the perfect choice.

The colours are subtle. Stylish. Sophisticated.

And every ingredient has been dermatologist tested for tolerance by even the most sensitive skins.

So you can wear RoC preparations with complete confidence.

Look elegant... and take better care of your skin. With safe, gentle RoC.

Getting the facts

If you think we have made a mistake in working out your supplementary benefit make sure you know how we worked it out.

If you don't understand the reason for the decision, or you've not had a form in the post showing how it was worked out, ask your social security office for it in writing. And look at leaflet SB.8 (or SB.9 if you are signing on) to help you to understand how we work it out.

The appeal letter

If you still disagree and wish to appeal, write to the social security office.

- You can use the form at the end of this leaflet or write a letter.

- You can get an addressed envelope from your post office.

- Say what you want to appeal about – you can appeal against any decision of the Supplementary Benefit Officer.

- Give all the reasons why you think the decision is wrong.

we're builders

Mr and Mrs Brick were builders.
They had been builders all their lives.
Their mothers and fathers
had been builders.
Their grandmothers and grandfathers
had been builders.
There had been builders
in the Brick family for years
and years and years.

Allan Ahlberg and Colin McNaughton

Chapter 1

It is a truth universally acknowledged, that a single man in possession of a good fortune must be in want of a wife.

However little known the feelings or views of such a man may be on his first entering a neighbourhood, this truth is so well fixed in the minds of the surrounding families, that he is considered as the rightful property of some one or other of their daughters.

Jane Austen

The National Criteria make a distinction between, on the one hand, writing:

in what may be termed 'closed' situations (e.g. the writing of letters, reports and instructions) where the subject matter, form, audience and purpose are largely 'given';

and, on the other:

'open' situations (e.g. narrative writing and imaginative/personal response to a range of stimuli and experience) where such factors are largely determined by the writer.

In the examples provided above, which ones do you think were produced by somebody who had been told in quite detailed terms what was expected of them ('closed' situations) and which ones were written by somebody who had decided what they were going to write about for themselves ('open' situations)?

If you are still finding it difficult to put your finger on what it is that makes one piece of writing different from another, think about what *Emma* and *Miss Brick the Builder's Baby* might have looked like if they had been written by somebody else.

Miss Brick the Builder's Baby
by Jane Austen

Mr Brick had never been more grateful for the inestimable good fortune that had accompanied him through a long and comfortable life, than in his decision, taken at an age when prudence is the least welcome of all the virtues, to marry Mrs Brick. She was a woman whose amiable disposition might not have provided such a compelling reason for their union were she not also able to count amongst her many personal accomplishments, a line of distinguished forbears who had prospered in an occupation not dissimilar to that of her husband.

Emma
by Allan Ahlberg and Colin McNaughton

Emma was happy.
Not only was she happy,
she was also clever.
Not only was she happy and clever,
she was also rich.
Not only was she happy and clever and rich,
she was also pretty.
Emma had been happy and clever and rich and pretty
for as long as she could remember.
And now she was twenty-one years old.

The differences between all those pieces of writing can be summed up under two main headings: 'Organisation' and 'Vocabulary'. Both of these headings are worth a closer look, so that you understand fully what you are expected to do when you sit down to write.

ORGANISATION

It's a wet Wednesday afternoon. You are sitting in a classroom with 28 other people. Some are scribbling feverishly away, covering sheets of paper at high speed, some are chewing their pens and staring out of the window. There's the occasional muttered conversation over in the corner, and somebody, hands in pockets, leaning back on his chair, seems to have finished already. You've been given an essay to do in order to round off a week of work on the effects of television. You've written down the title:

Some people say that television encourages violence. Do you think this is so? (MEG, 1986 Joint exam proposals)

and you've made a good start:

There are a lot of programmes on television that are violent— most of them are series about police work. I like watching programmes like this, and they don't make me violent.

Inspiration has dried up. You've got to fill another two and a half sides of paper . . . at least. An idea comes to you:

One of the most violent programmes I ever saw on television was a film in which a man got beaten up. They threatened him with a knife and when he wouldn't give them any money, they beat him up. There were pictures of blood coming out of his mouth and nose and ears when he was lying on the ground.

You are beginning to get into your stride now, and while you are writing, you remember part of a discussion that took place in class the week before. It was about whether violence on television encourages you to imitate what you see or allows you to get rid of your aggression by identifying with the characters on the screen.

> I think that if you watch violent television, you are less likely to go out and beat up old ladies because you don't need to any more because you have already had all the excitement.

Eventually, after a lot more agonising, you hand it in to the teacher — another week, another essay done.

If you go back and read that essay through again, ignoring the comments in between the sections, you'll see what the problem is. There are no connections between one paragraph and the next. Reading it is a bit like searching for a light-switch in a darkened room.

One of the most important parts of your job as a writer, then, is to devise the best way of organising the ideas and information that you want to get across, and remember, there are no simple formulae that you can apply to every writing task you are faced with. By looking at the way in which different kinds of writing are constructed, however, you can start to learn some general principles. Here is a newspaper article about a mysterious outbreak of Legionnaire's Disease in March 1985. It appeared in *The Observer*, but not in this form! All the paragraphs have been jumbled up like a jigsaw. Can you work out the order in which they originally appeared?

Cooling towers may have spread lethal bacteria

by OLIVIA TIMBS, ARTHUR OSMAN and ROBIN McKIE

1 Doctors changed their minds when younger people, two under 40 and nine under 50, began falling ill as well. Specialists from London's Centre for Communicable Diseases at Colindale were brought in and on Friday they identified Legionnaire's Disease.

2 High on their list of suspects are the giant power station cooling towers which dominate the Trent Valley round the town of Stafford.

3 Now local Tory MP Mr William Cash, is calling for a public inquiry into the outbreak and the local health authority's response to it.

4 A TEAM of Government scientists are searching this weekend for the source of the Legionnaire's Disease outbreak which has killed 28 people in Staffordshire — one of the worst such incidents on record.

5 Normally, the disease is confined to outbreaks in single locations where there have been infected air-conditioning systems or water supplies.

6 Some health officials suspect that insufficient chlorine, used to sterilise cooling tower ponds, may have been used at power stations, although electricity board officials dispute this.

7 Scientists believe vapour from towers at Rugeley and Stone may have spread Legionnaire's bacteria over much of the local countryside. Cases have been reported from homes in a 12-mile radius round Stafford.

8 The spread of the illness had originally led doctors to reject suggestions that Legionnaire's Disease was the cause of the outbreak of a virulent flu-like illness among elderly people in the area over the past two weeks. By Wednesday, 88 had been admitted to hospitals in Stafford, of whom 16 died.

9 'The position is very worrying,' he told *The Observer*. 'We have got to make sure we find out exactly what went wrong. Not least, we must find out why there was such a delay in finding out that Legionnaire's Disease was involved.'

11

Cooling towers may have spread lethal bacteria

by OLIVIA TIMBS, ARTHUR OSMAN and ROBIN McKIE

4 A TEAM of Government scientists are searching this weekend for the source of the Legionnaire's Disease outbreak which has killed 28 people in Staffordshire — one of the worst such incidents on record.

2 High on their list of suspects are the giant power station cooling towers which dominate the Trent Valley round the town of Stafford.

7 Scientists believe vapour from towers at Rugeley and Stone may have spread Legionnaire's bacteria over much of the local countryside. Cases have been reported from homes in a 12-mile radius round Stafford.

5 Normally the disease is confined to outbreaks in single locations where there have been infected air-conditioning systems or water supplies.

6 Some health officials suspect that insufficient chlorine, used to sterilise cooling tower ponds, may have been used at power stations, although electricity board officials dispute this.

8 The spread of the illness had originally led doctors to reject suggestions that Legionnaire's Disease was the cause of the outbreak of a virulent flu-like illness among elderly people in the area over the past two weeks. By Wednesday, 88 had been admitted to hospitals in Stafford, of whom 16 died.

1 Doctors changed their minds when younger people, two under 40 and nine under 50, began falling ill as well. Specialists from London's Centre for Communicable Diseases at Colindale were brought in and on Friday they identified Legionnaire's Disease.

3 Now local Tory MP Mr William Cash, is calling for a public inquiry into the outbreak and the local health authority's response to it.

9 ' The position is very worrying, he told The Observer. ' We have got to make sure we find out exactly what went wrong. Not least, we must find out why there was such a delay in finding out that Legionnaire's Disease was involved.'

Annotations:

This word provides the clue to the position of 2nd paragraph. It refers back to the scientists mentioned in the first 2 lines.

The key word here is 'normally' - the writers can only use it because they have already explained that this situation is *not* normal.

Taken together, these words provide evidence that you are well into the article - they assume that you know the basic facts already.

The word 'now' suggests that this paragraph has to come towards the end. The readers have been updated - they're *now* expected to think about the present situation.

You probably guessed correctly that this was the opening paragraph - it's the only one that gives you a general view of what the whole article is about.

It was probably difficult to spot this as the next paragraph, but you might have decided it should go in here because it provides more details about the cooling towers that have already been referred to.

Another difficult paragraph to place. The information was provided here because, as the article progresses, the facts are becoming more and more specific.

It should have been quite easy to link this paragraph with the one before it, even if you didn't place them correctly in the article as a whole. The writers are providing phrases that clearly signpost how this information fits the developing argument.

This has to go with the previous paragraph since 'he' could only refer to Mr William Cash.

If you look at page 12, you will be able to see what the article originally looked like.

Working through the newspaper article, you should be able to see that there is a logical order to the way in which information has been put across, and that key words and phrases have been used to help the reader understand what that order is.

Other kinds of writing require other kinds of organisation. If you are trying to write a story, you probably don't have to worry quite so much about the order in which you present things as long as you remember that you need a beginning, a middle and an end. However, you've still got to find solutions to other problems, like what to put in and what to leave out, how to get your characters from one situation to another, or what point of view you should tell the story from.

One way of understanding what this means in practice is to think about how you would tell a story if you were directing a film. A lot of books have a brief outline of the plot printed on the back cover. This is how the publisher describes *Beasts*, a book written by Dulan Barber for use in schools:

> When Carolyn takes a holiday job with the local research establishment, she knows she will be working with animals. But she doesn't know why the animals are there. Finding out gives her quite a shock.
>
> And when Carolyn's cousins, George and Gary, come across a mysterious man on the moors above the research laboratories, they do not think they will be drawn by him into helping Carolyn make a desperate protest against the way the scientists experiment on the animals.
>
> After all, is it right to use animals the way these ones are used? Why does it happen, and aren't there other ways of solving the research problems? All sorts of difficult questions face Carolyn and her cousins as they make up their minds about an important subject and then put their decisions into action.

That outline leaves a lot of questions unanswered about how the story is actually told. One sentence, however, is particularly important: 'Finding out gives her quite a shock.' If we follow this up, it gives us a chance to look at the kinds of problems that have to be solved by a writer of fiction. How can the author make sure that part of the story really works? The book describes Carolyn's arrival in the research establishment and her first meeting with Mr Benson, who is responsible for her. He shows her round and explains how she must look after the animals. Then he tells her not to open the door marked 'POST EXPERIMENT WING'. After a few days, however, Carolyn cannot resist. She goes through the door and sees animals in cages suffering horribly from the experiments. Very upset, she is found by Mr Benson and, bursting into tears, tries to run away.

1	2	3	4	5
Research plant in the distance. Carolyn walking up the hill.	Picture of lab inside research station.	Close-up of grim building – exterior.	Close-up of Mr Benson.	Carolyn alone with animals – showing affection for them.
6	**7**	**8**	**9**	**10**
Picture of door with sign: POST EXPERIMENT WING Door ajar.	Shot of animals in their hutches in the main animal house.	Carolyn working – cleaning out straw etc. from hutches.	Carolyn relaxing in lab with feet up having a tea break.	Picture of Mr Benson explaining how to do the work.
11	**12**	**13**	**14**	**15**
Picture of door with sign: POST EXPERIMENT WING	Shot of rabbit inside post experiment wing showing red, blind eyes.	Shot of dog in pain with goitre.	Mr Benson – alarm on his face.	Close-up of Carolyn's face – in tears.

You have been provided with fifteen scenes, all or some of which might be used by a director to turn this story into a film. However, they need *editing*. In other words, you have to decide:

which order they should be used in;
how long each shot should last compared with all the others;
which shots can be cut altogether.

You also have the freedom to decide whether you want to use just a part of a shot, in order to get a 'close-up' and whether you want to use the same shot twice.

Make a list of the scenes you would use, in the order you would use them, keeping a record of how long each shot would last.

When you are writing a story, you have to do all that, but instead of using film, you use words. Read on to see how Dulan Barber handled it in the book. How similar is this to the film that you would have made?

Mr Benson looked like Christopher Lee on a bad night: tall, gaunt and bloodless in a white coat. With an almost lugubrious slowness, he showed Carolyn where to park her bicycle, advising her to lock it, walked her to a stout chain-link fence, in which a narrow locked gateway was situated, and explained that she was never, ever to go beyond this point. Staring at the flat white buildings beyond, Carolyn said, quite truthfully, that she wouldn't want to. The Plant's silence still bothered her.

Then Mr Benson showed her where her locker was and gave her a clean coat and short white wellies. There was an electric kettle for making tea. And beyond this plain square room, stretching, it seemed to Carolyn, as far as the eye could see, was a long space filled with cages and pens. There were dogs and rabbits. White mice, rats of all hues, guinea-pigs, hamsters, cats and a romping litter of kittens. It was for this she had come, and the animals seemed to welcome her as she stopped to speak and pat, to rub soft fur with a finger eased between the bars. So busy was she making their responsive acquaintance that she barely heard what Mr Benson was saying about the hosepipes, the disinfectants, where the fresh straw and woodshavings were kept. He showed her storerooms, diet sheets, rosters and he could see that she was not taking it in.

'You'll soon get the hang of it,' he said almost with kindness. 'Ask me, or check the lists if you're not certain and I'm not about.'

Carolyn nodded.

There was a windowless flush door at the end of the room on to which had been stencilled, in stark black letters:

POST EXPERIMENT WING

'What's that?' she asked.

'Oh, no need to bother yourself with that, yet. You'll see in time. First get used to this lot.'

'OK.'

'You can make a start on the rabbits. Place them in an empty cage and thoroughly clean the dirty one, all right?'

'Yes.'

'I'll just stand and watch a while and see how you get on.'

Mr Benson shook his head and leaned against the wall, watching her. Sending him a girl! A good strong lad was what he needed. But Day, the local vet and inspector, had some pull with the top brass, so he'd had this lass wished on him. Wanted to be a vet herself or something and helping him was supposed to be good practice while she waited for her A-level results. Still, she was dressed sensibly and seemed willing enough. But he'd have preferred a strong lad, one he could let through that door without worrying. If there was one thing he couldn't abide, it was weeping women.

'All right,' he said. 'That's good. Carry on along the row. I'll be back in half an hour.'

'Thanks, Mr Benson.' Carolyn smiled and gently touched her cheek to the soft white, down-like fur of a large and quiescent rabbit.

The kittens had curled up, tired out at last, and Carolyn felt bored. It seemed hours since Mr Benson had taken the big black retriever bitch called Flossie out of her pen and through the gate in the chain-link fence. Carolyn had scrubbed and disinfected her pen and put down clean straw on the raised wooden platform. She'd taken the temperatures of the smaller animals and recorded them. She had, in fact, done everything Mr Benson had told her to do, and still he wasn't back. Perhaps she could tidy up one of the storerooms? She moved along the rows of pens and cages. There were certain times of day when the animals seemed to become listless or sleepy. There was no sound but the rustle of straw, the scutter of mice making phantom nests.

Suddenly, Carolyn was aware of the door marked POST EXPERIMENT WING.

She was curious. She hesitated. But, then, Mr Benson said she'd be working there soon. It couldn't do any harm to look, to familiarise herself. In fact, he'd probably be quite pleased at her keenness, initiative. Confidently, she opened the door.

There was a small dog, a brown and black mongrel, with appealing, golden eyes. It held its head on one side, as though permanently cocked, although it did not look alert. Carolyn stepped closer to its cage, peering. There was a fist-sized lump, a kind of goitre, on its neck. The dog, aware of a new, an alien presence, shuffled to the back of its pen. With a lurch of her stomach, Carolyn saw that the whole of its left side was encrusted with ugly mis-shapen lumps.

She turned away quickly, clutching her stomach.

The rabbit was blind, its eyes bulging, grey and red-rimmed.

Next to it lay a rat on its side, panting, panting, panting.

Another rabbit, almost bald, save for its head and long silky ears. Its pink flesh was horribly obscured by a sort of grey, web-like fungus.

'I thought I told you to keep out of here?' Mr Benson's voice scraped like chalk on a blackboard. The start it gave her, its sound made her burst into involuntary tears.

Carolyn ran towards him, tried to shoulder her way past. He caught her firmly by the shoulders and pulled her streaming face down to his white coat.

16

VOCABULARY

Two words that are commonly used to describe the way in which a piece of writing is organised are 'structure' and 'shape'. Besides having a 'shape', however, a piece of writing also needs to have an appropriate vocabulary; the author has to use the right kinds of words, bearing in mind both what is being written and who it is for. This is difficult, but not, perhaps, as difficult as you think. To some extent, in speech, it's something that we do instinctively. Take a look at the passage below, which illustrates the point very well. It's taken from a letter that was sent to a number of schools inviting students to send off £25 deposit in order to reserve a place on a sponsored tour of America. However, it was a hoax!

We are looking for 1000 young people who will be aged between 17 and 20 years of age on or before 1st August 1985, to take part in a *unique* sponsored two weeks trip to the United States of America *commencing* 16th August 1985. Included in the two week visit will be visits to New York, Chicago and Philadelphia.

The aim is for young people of Gt Britain to see for themselves the real America, rather than the image they see through the eyes of popular American television.

Various large American companies will be sponsoring the trip costing *well over* £600 000 therefore the cost to each applicant is *drastically* cut to only £50 per person for the two week stay. *Obviously* you will have to supply your own spending money and travel costs to and from London which will be the departure and arrival point, but all other expenses will be met by the organisers including all hotels, food, travel, insurance, organised tours etc.

The actual *itinerary* has still to be *finalised*, but the trip will include tours of large business premises, Universities, Museums, Art Galleries as well as various Leisure Centres and other places of interest. *Ample free* time will also be available for general sightseeing, all tours etc. will be *adequately* supervised by an *army* of tour supervisors.

The writers of the letter have chosen their words very carefully to create the effect they want. You can see how they've done it if you pay particular attention to the words in italics. The very first phrase: 'We are looking for . . . ' is intended to suggest that you are being singled out for special treatment. In the same sentence and for the same reason the trip is described as 'unique'. You're being flattered so subtly that you'd hardly notice unless you were looking for it. In the third line there's a word — 'commencing' — that's the first of many designed to make the trip sound official; 'starting' or 'beginning' would have done just as well, but they don't sound quite so formal. It happens again in the final paragraph, where words like 'itinerary' and 'finalised' are used. The impression is given that there's a big organisation behind the scheme, already beavering away on your behalf. Nothing, but nothing, could go wrong. In the third paragraph, they concentrate on the cost of the trip and the amount you will save. 'Well over £600 000' sounds a lot more than '£600 000', and notice that it's been cut 'drastically'. The generosity of the sponsors is also emphasised later on with the phrase 'Ample free time will be available'. What they're giving away, of course, is something you've already got, your own free time, but it sounds good. Finally, there's an interesting choice of words used to describe the tours themselves. They'll be supervised, the writers claim, by an 'army' of tour supervisors. What's being suggested, and this is probably for the benefit of teachers and parents, is that, like the army, the tour will be highly disciplined and no irresponsible behaviour will be tolerated. Very reassuring.

They've done quite a good job. It's only afterwards, when you realise it's a hoax, that you can see how it works.

SAMPLE WRITING

So far all the examples of writing that you've looked at have been produced by professionals – people like journalists, advertising copywriters, and novelists. Here is some writing by people of your own age (warts and all), trying hard to work in the kind of way that is needed for the GCSE examination.

Autobiography

To start, I must go back to my very early childhood. I remember standing in a doorway between the living room and kitchen. My mother was standing by the kitchen sink doing her washing. My mother was a small person, squat with dark black hair, short with a hard looking face, not ugly but not pretty either. She was always shouting throwing things. I was scared of my mum.

'fetch me a saucepan Dean'.

Like always I did not hear because she had her back to me. The next thing I knew she had lowered her self to my level and was screaming in my ear.

'I said fetch me a bloody saucepan'.

The high pitch screaming prevented me from hearing the whole sentence. I only ever manage to pick out one or two words. So I would just stand there and she would get very angry until in the end she would throw everything down, come storming into the living room, light a cigarette and smoke it. All the time muttering under her breath. After a bit she would get up and get on with whatever she was doing.

Even when my mother was like that I loved her very much, maybe because I did not have a father. My mother and father were divorced when I was two. Seeing as she was all I could have held on to. I did. I have a sister a year older than me. We did not get on because my mother always had time for her, but never for me. I guess I felt left out and blamed my sister for that.

Apart from my mother and sister, I very rarely saw anybody else except my nan. She would come to our house on Wednesdays. I would go down to her house on Saturdays. She was a nice old lady, big all over and had a very understanding nature. She had some very rare qualities like she was honest, sincere, genuine, about the nicest person I have ever met. My grandfather is a large man with white hair and is a very quiet person. A strange man, very rarely talks, just listens to his grammerphone, plays chess or reads books. I did not like him as much as I liked my nan because he was not as frienldly. He would come into the kitchen on a Saturday night, just after we had had our tea, give us a devon toffee and smile at us. Sometimes he would make a funny comment. I never understand what he meant. My gran always laughed, so I did too.

By now I was old enough for school. I did not want to go but my mother made me. The school was a huge grey building. I did not like it at all. There was a lot of children running around laughing and screaming. I grabed hold of my mother's coat, a big red fluffy thing with great big red buttons. I would not let go but was made to. I was taken down a corridor by an old lady of slim build. She was wearing a very tight skirt which came just below her knee. She had a jacket to match. It was made of a very thick woolen cloth. The school felt clean like a hospital. She took me into a class full of childern. I had stopped crying by now and looked around. I was told to sit down, so I walked up a row of desks and sat down next to a boy with a dark complexion. He did not seem scared like me. He was looking around, happly laughing and grinning. I liked him a lot, he put me at my ease.

If you recall the 'assessment objectives', you'll remember that one of the things you have to be able to do is to 'articulate experience and express what is felt and what is imagined'. That's exactly what Dean is trying to do in this piece, which is only the beginning of a much longer 'autobiography' covering his whole life up to the age of 16.

How well has he done it? Before reading on, try to put yourself in the position of an examiner, and think how you would grade a piece of writing like that.

The first thing to notice is that the organisation is very simple, but quite logical and it works surprisingly well. There's no beating about the bush. 'To start, I must go back to my very early childhood,' he writes, and then he launches into a description of a particular incident that he remembers very vividly. Although he doesn't actually say so, it's a fair bet that this is, in fact, his first memory. Don't be misled by the simplicity of it, however, into thinking that he's chosen to start here for that reason alone. He makes it quite clear, from the words he uses, that this incident is intended to convey a more general impression of what things were like at home when he was young. 'She was *always* shouting throwing things,' he writes, and '*Like always* I did not hear'. That makes the opening work much more successfully.

The second paragraph starts with one of those linking phrases that helps readers find their way through a piece of writing: 'Even when my mother was like that . . .' He's clearly telling us that he's now going to *comment* on the scene he's just *described*, and that's exactly what he does. He tries to explain why he loved his mother, even though she wasn't easy to live with.

Both the third and fourth paragraphs are equally well signposted. The third one starts: 'Apart from my mother and sister . . .'. The message is plain. He's told us about the things that mattered most to him as a child, now he's moving on to write about his 'Nan'. Having done that, we've reached school age. You might protest that he's not spent much time on the first five years of his life, but at least the sentence at the beginning of the fourth paragraph − 'By now, I was old enough for school' − leaves you in no doubt where you are, and links in very smoothly with what has gone before.

The *organisation*, then, has been given quite a lot of thought − he didn't just sit down and start writing. And what's most important about it is that he's always kept his reader in mind − remember what the GCSE criteria had to say about 'audiences'? He's never forgotten that somebody's reading this who doesn't know him, and so he's good at judging how to give you the necessary information. Look at the opening of the second paragraph again. After explaining that he was attached to his mother because he 'did not have a father', he realises that this needs some *explanation* − 'My mother and father were divorced when I was two.'

What about what he has to say and the words he chooses to say it in − the *vocabulary* we mentioned before?

An examiner would immediately notice that the language is all fairly simple. There are no lengthy descriptions and most of the sentences are very straightforward. However, the words he has chosen to describe his mother's behaviour work well because they seem to have been picked very carefully. Notice that she comes 'storming' into the room, for example, and that she is 'muttering' to herself. There are also other details that he mentions, like the 'devon toffee', which are very precise and bring the scene to life. The moral of this is that careful observation can sometimes produce more successful description than any amount of adjectives ladled over an essay like custard.

Left until last is the most important point that needs making about this piece of writing. That's because it's also the most difficult to explain. Although the language and the organisation are all *simple*, the feelings that he is trying to convey are *not* — they are as complicated as real life usually is. One example of this has been noted already — his feelings about his mother. But there are plenty of other places where you can see it. He has difficulty describing his mother's appearance — 'not ugly but not pretty either', he says. And his grandfather is similarly difficult to pin down — 'A strange man, very rarely talks, just listens to his grammerphone, plays chess or reads books.' These, you feel, are real people, viewed with absolute honesty. He achieves this because he doesn't tell the story as a child would but as an adult looking back on childhood — 'Sometimes he would make a funny comment. I never understand what he meant. My gran always laughed, so I did too.'

A successful piece of writing then? Well yes, up to a point. The assessment objectives also say that candidates must 'exercise control of appropriate grammatical structures, conventions of paragraphing, sentence structure, punctuation and spelling'. And Dean clearly has a bit of a problem there. Unfortunately, some words are misspelt: 'grammerphone', 'grabed', 'childern', 'happly'. On the other hand, and with a bit of luck, the examiner might notice that he can spell 'complexion' properly, which you might have expected to cause him some difficulty. There are also sentences that aren't proper sentences, and commas where there should be full stops. All of these things would count against him in the final mark, but, since they don't interfere too greatly with his ability to communicate what he has to say, he'd probably still get credit for the good things about his essay. Have some sympathy for the examiner, though, since it's not easy to compare all these different features of a piece of writing and arrive at a grade that fairly represents all of them.

That's an example of just one kind of writing amongst the many that you might have to turn your hand to. The second example is very different. It fits best the assessment objectives in the National Criteria that refer to 'conveying information' and 'ordering and presenting facts, ideas and opinions'. You'll notice that some of the comments made about it are similar to those made about Dean's autobiography, whilst some are based on different criteria altogether. But first read this extract from Paul's 'Guide to Salt Water Fishing'. It's just one section of the whole guide and it deals with the 'baits' that anglers use.

Baits

There are five main types of sea bait used to catch fish in British Waters. These are; Lugworms; Ragworms; Crabs; Mussels, and strips of fish. This bait can be found along most coastlines with the exception of Rocky Coasts. If for some reason the bait cannot be found along the beaches, it can be ordered from most of the better Tackle Shops.

The lugworm is the commonest of the five, and lives on beaches and estuaries. The worm grows anything up to about seven inches in length. To look at the lugworm resembles the earth worm; but the main differences are that the lugworm is larger in diameter and has an Extension 'Tail'. These creatures can be found in 'U' shape tunnels in the sand. Although the tunnels cannot be seen from the beach the small heap of defecated sand, the worm clears from the tunnel, can be seen. The 'tail' contains nothing except sand and is usually removed by the angler. The juices of the worm although may seem unpleasant are the main attraction for the fish.

The ragworm is a hideous looking creature and digging for them is a dirty job. The ragworm can grow up to fifteen inches in length and looks like a giant Centipede. They can be found in mud flats and estuaries. To keep the worm fresh when caught, you should fill a wooden box with sea weed. The worms when placed into this will keep fresh for up to two days. The ragworm has a reputation for having a nasty bite. Do not be put off by this nonsense as the 'nip' only resembles a pin prick. The ragworm is the best bait for bottom feeding fish.

Little really needs to be said about crabs. They can be found in harbours, beaches and rock pools at low tide. You should use crabs with a shell length of one to four inches. The soft back crab is simply a hard backed crab that has shed its shell. The crab should be used in a live state and can be kept for two days in a seaweeded box. The mussel is a very common form of bait, and is an excellent alternative to those mentioned above. The creature should be removed from its shell before being placed onto the hook. The mussel is the best bait for cod and haddock. They can be kept for longer than twenty four hours if the water is changed. When hooking the mussel, it is advisable to hook the orange coloured 'foot', as the rest of the body is too soft to hold a hook.

Paul's aim in this is quite clear. He wants to give sound advice to beginners about how to become expert sea fishermen. The main problem that the task presents him with is equally clear. He's got to find a way of presenting his wealth of knowledge and experience in such a way that it can be understood by somebody who knows next to nothing about it. That means he can't use technical terms without explaining them, he's got to remember to refer to all the basic things that he probably does without even thinking about them, and he's got to inspire confidence in the readers (does he really know what he's talking about?), whilst also reassuring and encouraging them. A pretty tall order.

Nonetheless, the examiner will have all those points running through his head as he reads and tries to judge the guide as a piece of writing.

There's no doubt that Paul has got a very organised and logical mind. The first paragraph lists the 'five main types of sea bait' and the succeeding paragraphs deal with them one at a time. At least, that's the theory. In practice, crabs and mussels are put together and 'strips of fish' don't actually get a mention. That's a weakness, but not too serious since he clearly feels that they don't need to be explained in quite so much detail – 'Little really needs to be said about crabs.' The reader is still able to follow without any difficulty. Paul manages to make it all seem clear and simple to a novice. Once again, though, don't be fooled into thinking that he's managed to do this so successfully because he's chosen to write about something that isn't actually very complicated. The word 'main' in the first sentence suggests that there are other 'types of sea bait' and that he's had to go through that difficult process of *selecting* the information that is most relevant to the job.

Selecting and ordering the information is only part of the task, however. The examiner will also be considering whether Paul has chosen his words carefully. If questioned about this, the examiner would probably say that he or she was judging the 'tone' that Paul had adopted.

The first impression is of somebody who doesn't suffer fools gladly. Paul doesn't allow his readers much time for understanding before they are rushed on to the next point. You feel that if he was actually teaching you, he might lose patience if you got things wrong the first time.

A second reading, however, does something to correct this. He's actually a bit more sympathetic than he sometimes sounds, and half way through he even addresses the reader directly – 'You should use crabs with a shell length of one to four inches' – which helps. In addition to this, he's very thoughtful in imagining possible problems and providing good down-to-earth advice: 'If for some reason the bait cannot be found along the beaches, it can be ordered from most of the better Tackle Shops.' Probably the phrase that makes the biggest difference, though, comes at the beginning of the paragraph about ragworms: 'The ragworm is a hideous looking creature and digging for them is a dirty job.' Suddenly, you appreciate that he knows what he's talking about because he's done it. His personality comes across despite the fact that, most of the time, he rightly adopts a very unemotional tone.

However, the acid test for a guide of this kind is whether he has managed to explain things clearly. Take this, for example:

> The ragworm has a reputation for having a nasty bite. Do not be put off by this nonsense as the 'nip' only resembles a pin prick.

He could have got into a real mess trying to explain this point, but he uses two short sentences and gets his meaning across very clearly. The key phrases are 'has a reputation' and 'resembles'. Both enable him to use a single word where other, less accomplished writers might have had to put in a whole sentence. Try re-writing these sentences to make them briefer. It's virtually impossible without losing some of the meaning, and that's often a good test for a piece of writing of this kind.

The whole of this section has been aiming to take you through the way in which an examiner, or, indeed, a teacher, might look at your work. It's been quite complicated, but if you've understood, you'll already be part of the way towards becoming a better writer yourself. Neither Dean nor Paul think writing is easy, and they'll both find it hard to get top marks. But they've made the most of the talents they do possess by developing good writing habits, and that is what the next section is designed to help you achieve.

GOOD WRITING HABITS

Each part of this section is introduced by one of the questions that you need to ask yourself if you are going to improve your ability to write.

What am I going to write about?

It may not feel like it, but even when you've been given just one essay title by a teacher, you still have a great deal of choice about what you are going to write. And a lot of the time, you will find yourself having to pick one question from amongst many. Here's a typical examination paper:

Section A – Continuous Writing (30%)

Write on one *of the following subjects. You may include a plan to help you construct and organise your writing. It will* not *be marked.*

You should write between 350 and 600 words. Credit will be given for relevant and well-ordered ideas, use of appropriate language, careful spelling and punctuation.

1 **A job well done**

Write about a day when you tackled something that took time, thought and effort. Describe the task in detail and your thoughts and feelings.

2 **The view**

Imagine that you have climbed to the top of

either   a cliff overlooking a stretch of coastline;

or   a hill overlooking a small, industrial town.

Describe carefully the climb and what you saw on the way and from the top.

3 Video cameras mean that many people will be able to make their own movies on cheap tape instead of expensive film. Imagine that you have a camera and recorder and you decide to make your own film. Tell the story of how you make the film, bearing in mind that things do not always go right and that producers need a sense of humour.

4 **Training pets is never easy**

5 'I wish people, when you sit near them, wouldn't think it necessary to make conversation and send draughts of cold words blowing down your neck and in your ears and giving you a cold in your inside.'

Either   Write an essay on 'Conversations I would rather not have had'.

Or   Invent, in play script, *two* conversations you can imagine yourself having: *one* that you would enjoy and *one* that you would not.

6 **A day to remember**

7 **A disused canal, railway line or building**

Describe *either* a canal, *or* a railway line, *or* a building so as to suggest its disused and derelict appearance. You might wish to use part of your writing to compare the present appearance with what you think it might have looked like when it was busy and in use.

8 **The advantages and disadvantages of being a 'teenager'**

9 In the course of a journey, you meet a most unpleasant person. Describe the rest of the journey.

10 Write in any way suggested to you by *one* of the pictures on the accompanying sheet.

In the exam room, you'll have to make a fairly rapid decision about which of these questions you want to do, and if you make the wrong choice you could waste a lot of valuable time. It's important, then, that you are able to judge what *kind of writing* each question demands, and that you *know your own strengths and weaknesses*.

Here's a checklist that might help. Are you good at:

	1	2	3	4	5	6	7	8	9	10
telling a story									✓	
creating characters									✓	
describing people and places		✓							✓	
inventing conversation									✓	
constructing arguments				✓						
conveying facts and information				✓						
expressing your opinions				✓						
writing about feelings		✓								
imagining events		✓							✓	
being persuasive										

The left-hand column asks the questions that you must eventually learn to ask yourself. The numbers one to ten along the top refer to the essay titles on the exam paper, and three of them have been filled in with ticks showing what you'd need to be able to do in order to get high marks in the exam. Do you agree? Could you fill in the empty columns in the same way?

Sometimes the way the question is worded will tell you all you need to know. Take number **2** for instance, 'The view'. It starts with the instruction to '*imagine* that you have climbed to the top of a cliff . . .' and goes on to ask you to '*describe carefully* . . .'. Question **4**, on the other hand, 'Training pets is never easy' isn't quite so helpful. Because there is just the title by itself, you have more freedom to interpret it the way you want. However, the way it's phrased suggests that the examiner isn't looking for a story, say, or a description of your pet but is expecting you to write about pets *in general*. If you chose this question, you'd have to be able to *explain* what is involved in training pets and give an *opinion* about how difficult it is, perhaps pointing out some of the problems that you might be likely to come across.

The questions on that examination paper are all of a fairly familiar kind. However, if you are going to submit a coursework folder instead of sitting an exam, you could find that you are expected to approach the whole business of writing in a very different way. You might, from time to time, be asked to answer the tricky question with which this section opened, 'What am I going to write about?' for yourself.

Faced with that, even the most experienced writer can sometimes be thrown into a panic. Your mind goes as blank as the clean sheet of paper in front of you. The first thing to remember is that you need to find a subject on which you can be an expert. Look at the following titles:

The story of my life

A guide to buying and keeping tropical fish

Part-time jobs locally for 16 year olds: a survey

The Inventor Stories: written for children at St Thomas' Primary School

My Grandad: wartime memories

Keep your bike on the road: an instruction manual for owners

Change the wording around a bit and these are all ideas that you could try out for yourself. The thing about coursework is that it allows you to be a bit more ambitious, and to write something that is personally more satisfying to you. Make the most of that opportunity.

SUMMARY

1 Judge what kind of writing each question demands.
2 Know your own strengths and weaknesses.
3 Make sure that, whatever you are writing about, you are the expert.

Where do I start?

If you were in a workshop about to make, let's say, a table, you'd probably start by gathering together all the things that you needed to do the job: wood, glue, screws, saw, sandpaper, and so on. It's the same with writing. You need to assemble all the ideas, information and scraps of research that might come in handy. And you need to do this well before you actually start writing.

However, there are no hard and fast rules to tell you how to go about it, so you need to be flexible enough to choose a system that suits both you and the essay that you are working on. Some examples might help to show you a few different approaches.

The most usual way of planning a piece of writing, and the system that you have probably been taught to use in school, involves preparing a 'skeleton essay'. It's a 'skeleton essay' because all you put down on paper are the bare bones of what you want to say, often as a series of numbered points. You never use more than a few words to remind you about what needs to go into each section or paragraph and, because it's only a rough outline, you can go on adding ideas right up to the point where you are ready to start writing. Let's go back to that question on the exam paper about describing how it feels to climb to the top of a cliff overlooking a stretch of coastline. Your skeleton essay might start off something like this:

1 Opening - climbing out of the seaside resort, ← ⎡Description of family - all shouting at
  leaving day-trippers behind.   ⎢each other, kids complaining, dropping
                                                  ⎣sweet-papers, etc.

2 Walking up the hill, getting tired, stopping and
  looking around.

3 Part of the path has slipped and looks dangerous -
  trying to find a way round.

4 Meeting a farmer who doesn't like me being on ← ⎡Farmer very hostile - conversation - explain
  his land.   ⎣that path has slipped so can't go back.

5 Arriving at the top exhausted - looking down, ← ⎡Everything very small. How quiet it is at the
  the view.   ⎢top. Feel of lying on my stomach on the
                                                  ⎣grass.

Having got this far, you might feel ready to start straight away, but it's worth being patient and re-reading what you have written to see whether you have any other ideas that need to be fitted in. In the second column above are some additional points of the kind that you might jot down at this stage.

Only when you've done this should you actually begin your essay. And remember, your plan isn't sacred. If, when you are actually writing, it doesn't seem to be working out as you expected, be prepared to make changes.

That's a simple and effective way of working out your ideas which doesn't take up too much time. However, it's a system that works best when you don't have to think too hard about the order in which events happen. With some kinds of writing, this can be much more of a problem than in 'The view', and you may find that you need to develop a different way of working. Take another question, 'The advantages and disadvantages of being a "teenager" ', for example. There's no obvious starting-point and no obvious way of seeing what you should put in or leave out. Some people find it easier to plan an essay of this kind by using a diagram to help them, as this provides a way of delaying the problem of what goes into the opening paragraph until all the material has been collected together. They are then free to *brainstorm*. In other words, they can jot down everything they can think of that might be in the slightest bit relevant, however stupid it may seem, secure in the knowledge that they can chuck out the bad ideas at a later stage. Whatever kind of diagram you draw, the basic principle is the same: you draw lines across the page to connect ideas with each other. Like this 'topic web':

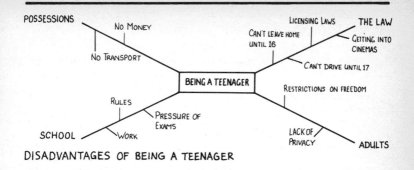

DISADVANTAGES OF BEING A TEENAGER

The final example might give you some ideas about how to start a piece of writing for a coursework folder, when you have the opportunity to do some research beforehand. Not all research has to be done from books. You might decide, for example, to write a *biography*, to tell the story of someone else's life. Here is an extract from a worksheet that one school uses in order to give advice on how to set about a piece of work of this kind.

Having decided who you are going to write about, you need to <u>talk</u> to them and find out as much as you can about them so that you can put yourself in their shoes. You need to be able to imagine what they must have gone through, and what they must have felt and thought.

You can either take notes or use a tape recorder.

*Advantages of using a tape-recorder

1.   It is easy to concentrate on what is being said and think up new questions.
2.   You have a complete record of what has been said.

<u>Disadvantages</u>

1.   Some people get nervous when a microphone is pointed at them.
2.   It is difficult to find the right place on a tape when you are referring back to it.

*Advantages of taking notes

1. The person speaking doesn't feel inhibited.
2. Easy to refer back to.

Disadvantages

1. You can't listen and question very easily.
2. You can't get everything down on paper unless you can write very fast indeed, or have lots of pauses.

You can borrow a tape-recorder from the English department. If you intend to use one, remember to hold the microphone close to the person you are talking to. They will quickly get used to it and you will have a clear recording.

*What do you need to ask?

This is a short passage from a book called 'Millstone Grit'. An old lady is talking to the author about her life and he has recorded and written down what she has said:

"I was born at a farm up Soyland. My father had a lot of bad luck. He wasn't farming then, he was a paper finisher - they used to count the reels as they came from the mill. He kept going to a job and then something happened so he went to another one and we were moved around quite a bit. In those days a lot of mills were burnt down. That at Ripponden was burnt. When a mill burnt down they couldn't stand around in the village, they had to walk somewhere and get something else. It happened quite a few times and I suppose he got a bit frustrated and he went to America when I was about thirteen. He worked his passage across and only stayed about twelve months."

There are lots of interesting things in this - Mills burning down, a trip to America, but it is difficult to know from what she says how they affected the old lady. If you had been doing the interviewing with a view to writing, you might have interrupted and asked:

> Why were the mills burnt down?
> What other jobs did the men get?
> What did your father do in America?
> Did you ever see a mill burning?
> Can you describe it?
> When did you discover your father was going to America?
> What did you think?
> Can you remember saying goodbye?
> Can you remember the homecoming?

All these questions would help you to imagine the scene well enough to be able to describe it for somebody else.

If you can track down somebody with an interesting fund of stories to tell, and persuade them to talk to you, you could find yourself with a great deal of information that can then be organised by using a skeleton essay or a topic web.

If this all seems a bit daunting, the best advice is to divide your writing into sections (sometimes, as with a guide or instruction manual, you can even write the headings in) and make sure that you cover each section fully before moving on to the next. That way a task which seems impossible can be reduced to manageable proportions.

SUMMARY

Where do I start?

1 *If you are doing coursework, research your topic fully.*
2 *Plan your writing by making:*
   (a) a skeleton essay, or (b) a topic web.
3 *Brainstorm ideas before selecting the best ones.*
4 *Think in sections.*
5 *Complete each section fully before moving on to the next.*

How do I finish?

The main difference between sitting an exam and producing a coursework folder lies in the answer to this question. If you are sitting an exam, you won't have time to re-read your work and alter it. You have to get it right first time, and the chances of doing that are so remote that you'll never be entirely happy with what you've written. It's a bit like asking somebody to cook a meal without allowing them to taste any of the ingredients as they go along. No professional writer would dream of working under such conditions.

It's quite possible, however, that many candidates will continue to be assessed in this way. If you happen to be one of them, try and find time to re-read your work and, neatly, make any changes that are necessary. You can also help yourself by looking back over what you have written just before you tackle your final paragraph. That allows you to provide a conclusion that makes quite clear what you intended to say, even if the essay didn't turn out exactly the way you wanted it.

The real opportunity that GCSE offers, however, is for you to become expert at revising your own work without being under pressure from shortage of time. The National Criteria put it this way: 'Coursework . . . offers more realistic conditions for drafting and re-drafting'. This is not as easy as it might seem. First of all, you need to be very careful about what counts as cheating. Exam boards will have their own regulations which you should be aware of. In general, though, you should not re-write an essay that has already been marked by your teacher. That doesn't mean that the teacher cannot give you any help at all — simply that the discussions you have must take place before the production of your final version.

That, however, is just a matter of sticking to the rules, and you'll find that teachers and examiners can tell very easily whether something is all your own work. Much more difficult is developing the ability to step outside what you are writing in order to see how it might look to somebody else, yet that plays a vital part in the process of deciding whether to alter a word here or change a sentence there. The stages through which this book has gone before it reaches you provide a perfect example of this point. As each page is prepared for typing, it turns into a mass of squiggly lines, crossings out, alterations and additions. Everything has had to be read and re-read in order to gauge whether it can be understood by somebody who isn't familiar with the business of setting and marking exams.

There are ways in which you can help yourself become more effective at doing this. When you have written something, put it on one side and come back to it a day or two later. The gap between writing and reading will help you to see it with a fresh eye, and sometimes you'll find that there are whole sentences that seem to be so confused that you can't even remember what you were getting at. It's also worth reading your work aloud, to a friend or to your parents. This can be excruciatingly embarrassing, but only because, as you are reading, you became aware of what is wrong. Somehow the knowledge that some-body else is listening helps to distance you from what you have written. Finally, if you have produced a piece of work that has a particular audience in mind, try it out on them and get their reactions. If it's a story for a seven year old, a seven year old may well be the best person to tell you how to make improvements. If you've produced a guide to a walk in your local neighbourhood, then ask a friend to use it. If he or she gets lost, you'll know you need to try again.

If you start off in your best handwriting, thinking you can save time by producing a fair copy first time off, then you'll never re-write. It will all seem too much like hard work. Decide, right from the beginning, that you are only trying out ideas, and it doesn't matter if you scribble, or get coffee stains on the paper, or cross things out. That will leave you free to alter things, and give you the pleasure of seeing it all fall neatly into place when you complete your final version.

Here's an exercise designed to help you get your hand in. Imagine you are working in a newspaper office. Reports have been phoned in to you about a disturbance caused by a motorcycle gang in Tynmouth, a Cornish seaside resort. Sticking closely to the facts as they reached you, you have written the following brief article:

Holiday Horror

The bank holiday calm of Tynmouth was rudely shattered yester-day when motorcycle gangs from nearby towns descended on the sleepy Cornish resort. Large numbers of motorcyclists were able to by-pass police checkpoints on the approaching roads and gather in the town centre. Terrified residents watched in horror as the gang rampaged down the main street when police attempted to cordon off the area, and traffic was brought to a standstill. Order was finally restored when the motorcylists were escorted towards the sea front, though scuffles broke out and several arrests were made. Nervous holiday-makers stood their ground on nearby beaches refusing to be intimidated.

However, just before you sent your 'copy' to the editor, you received the following additional information:

Martin Stevens, aged 9, who was swimming off the coast at Tynmouth, was in danger of being swept out to sea by strong cur-rents when Peter Dixon (aged 22) and Terry Strain (aged 18), both members of the motorcyle gang that had gathered on the beach, dived in and made a heroic rescue attempt. Stevens is now in hospital, but is likely to be discharged tomorrow.

How would you rewrite the article?

How do I finish?

1 *Get into the habit of re-drafting your work thoroughly.*
2 *Never start with the intention of producing a fair copy at your first attempt.*
3 *When re-drafting, make an effort to imagine how your work will appear to a reader.*
4 *Techniques to help you achieve this include:*
   (a) re-reading a day or two after you have written;
   (b) reading your work aloud to a friend;
   (c) testing your work out on the audience for which it was intended.

How should my work be presented?

Presentation could make a significant difference to your final grade. Research has been done, for example, which shows that teachers and examiners are unconsciously affected by poor handwriting, even though they try hard not to let it influence them.

However, it's coursework assessment that is likely to make the greatest demands on you. At the end of the course, your teacher will have to gather together a sample of your work, including some pieces that may have been written a year earlier. There may be all kinds of regulations about how many examples of different kinds of writing you will need to submit, and you'll certainly be expected to provide information about when a piece was written and what task you were set. That means you can't afford simply to stuff your work in a drawer and forget about it. You've got to be organised.

First of all, you should check on exactly what the requirements for the coursework folder are. This, for example, is taken from a GCSE syllabus produced by the Southern Examining Group:

Paper 3 – Coursework   *50% of the total marks.*

The coursework folder must contain 7 pieces of work of an approximate total length of 3000 words. A variety of writing is required: at least two of the pieces being written in response to literature read during the course and at least one of each of the following:

   (i)   a description
   (ii) a narrative or a piece of creative writing
   (iii) an explanation
   (iv) a report
   (v)   an argument or a piece of discursive writing

It is expected that the pieces chosen will represent the candidate's best work in each of the above categories. The folder should not be produced under formal examination conditions but under conditions which will enable centres to guarantee its authenticity. Pieces may be written in class or at home, with or without prior discussion with the teacher and a description of the conditions under which each piece is written must appear on the coursework mark-sheet. However, two of the seven pieces must be written in class under supervision. Coursework folders should normally contain pieces written during the academic year in which the examination is taken, but some pieces may be included from the previous year's work if desired by the candidate. Each piece should be marked when it is completed and the entire folder must be available for submission for moderation by a date to be notified.

Do not assume that every syllabus will be like this one. In your area, the exam board may want something different, and your teacher will certainly be making sure that the right kind of work is set during the course. However, you should still get to know what is expected of you, so that if you find, for example, that all your best work falls into one category, 'description' say, then you can do something to bring the rest of your folder up to the same standard.

You may be used to doing all your work in exercise books, and the change to writing on file paper could cause you some problems, particularly if you are in the habit of losing things. You would be well advised to keep a folder or a file specially for your English coursework and put everything in it. If you feel confident about distinguishing between different kinds of writing, then you might try using dividers in order to file your work under each of the headings that are required:

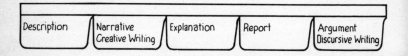

Description / Narrative Creative Writing / Explanation / Report / Argument Discursive Writing

Finally, you must remember to put proper headings on your work. Some of the information is needed for reasons that are obvious. Teachers, examiners and the Post Office are just as capable of losing work as you are, and you need to do everything to ensure that if your folder goes missing, or falls apart, then your work can be quickly identified and reassembled. You may, however, be wondering why you need to date your work. The answer is simple. Although you are unlikely to get any credit simply for making progress, an examiner might not mark you down for poor work done early in the course if the final pieces are good enough to get a higher grade.

SUMMARY

How should my work be presented?

1 *Poor handwriting could lose you marks.*
2 *Check on the exam board's requirements for your coursework folder.*
3 *Keep a folder specially for your English work and store it in a safe place.*
4 *Use dividers to file your work according to the different kinds of writing required by the exam board.*
5 *Provide full and accurate headings for your work.*

SECTION 4

Reading

INTRODUCTION

In both 'O' level and CSE exams, your skill at reading was tested by what became known as the 'Comprehension', in other words, a short passage accompanied by a series of questions designed to find out whether you had understood what you had read. The trouble was that schools tended to train pupils not to become better at reading, but to become better at answering comprehensions — and there is a world of difference between the two. For the GCSE some of the exam boards have dropped the term 'comprehension' altogether. If you live in the Midlands, for example, it's possible that you'll have to take an exam paper called 'Directed response to expressive writing' and in London there's a whole section of the syllabus devoted to 'Understanding and response'.

The other point they're trying to make is this. Reading is something that you do for a whole variety of reasons. Whether you are poring over the back of a prescription in order to sort out if you qualify for free medicine, or settling down with a book to entertain yourself for a few hours, you have a *purpose* that influences what you read and the way you read it. Tests of comprehension, on the other hand, exist in a world of their own. The only reason there can be for doing them is to demonstrate that you have understood them. That's it. It's as if you read books just in order to show that you were able to.

A useful way of grasping this idea, that the reason you have for reading alters the way in which you do it, is to think about recipes. Here's one for Steak and Kidney Hotpot. Before you look at it, imagine that you are flipping through a cookery book trying to decide what to choose for supper. How do you read it? Do you take in every word, starting at the beginning and working through to the end, or do you just glance at the description and check the contents quickly to decide whether there's anything there you don't like?

Steak and Kidney Hotpot   (Serves 6)

If you love the delicious combination of steak and kidney but aren't too keen on suet pastry, then this recipe is for you, because instead of pastry it has a crust of thickly sliced potatoes on top.

1½ lb chuck steak, trimmed and cut into bite-sized cubes
½ lb ox kidney, trimmed and cut fairly small
2 lb potatoes, peeled and cut into thick slices
2 medium onions, peeled and roughly chopped
1 rounded tablespoon flour
½ pint beef stock
¾ teaspoon Worcestershire sauce
Beef dripping
Melted butter
Salt and freshly milled black pepper
Pre-heat the oven to mark 2/300 °F

First melt some beef dripping in a large, wide-based saucepan and fry the onion in it to soften for about 5 minutes or so; then turn the heat right up, add the cubes of beef and kidney and cook them to a nutty brown colour – keep stirring and turning the meat as it browns. Now lower the heat a bit, sprinkle in the flour and stir it around to soak up the meat juices. Season well, add the Worcestershire sauce, then gradually stir in the stock and bring to simmering point. Next pour the meat mixture into a casserole or pie dish and arrange the thickly sliced potatoes in layers all over the meat. Season the potatoes, brush them with melted butter, then cover the casserole with a lid or foil and bake in the oven for 2½–3 hours. Before serving remove the lid and brown the potatoes under a very hot grill to get them really crisp.

Now go back to the beginning. This time, think about how you'd read it if, having decided that this is the meal you want, you had to make up a shopping list.

Finally, try to sort out what you'd make of it if you were actually in a kitchen, cooking.

There's one general point to be made. Each time you looked at that recipe, you probably started somewhere different and read different words. That's part of what reading is all about, getting better at finding the most efficient way of using the printed word for your own particular purposes.

That example, however, doesn't tell the whole story. By and large, you can trust a recipe. Nobody's going to use a cookery book to fool you into doing something you don't want to, or believing something that isn't true. But that's not the case with everything you read. Good readers have to be good detectives, able to look at the words carefully enough to spot what's really going on.

You can get some idea of what this means by reading the eye-witness accounts printed below. They give you three different versions of what went on during a fight at a disco. By comparing the three accounts, you should be able to work out who is most likely to be telling the truth.

Peter, after being arrested by the police for fighting, provided this statement:

I was standing at the edge of the disco, away from the doors, close to the steps that lead up to the raised area where the bar is. The fight started quite near me, and I couldn't get out, so I escaped by going up the steps. I watched from up there for a few minutes until I heard the sirens going as the police arrived. By then, quite a lot of people had joined in the fight, but most of them had been bundled out of the main doors. I thought it was safe to try and get out, so I went down the steps and made my way along the wall keeping out of trouble. A bouncer just came towards me, picked me up and threw me back against the wall. I was so dazed I couldn't do anything. He grabbed me by the arms, dragged me to the doors and threw me on to the pavement. The next thing I knew I was in a police van. I didn't start anything.

The bouncer has a different story to tell:

I was up at the bar talking to the barman when the trouble started. I immediately went down the steps to the fight. Nobody passed me, all the kids were crowding round to see what was happening. I had just started to clear the area, when this idiot came running towards the fight, fists flying. It's only because of me he didn't get hurt more badly, since one or two people had been tempted into having a go at him. That's how he got his injuries. I restrained him without using unreasonable force and took him to the main exit where I let him go and told him to clear off. Then I went back to help sort out the mess.

The policeman's account is as follows:

It took a while for us to reach the disco as it was closing time when we got the message and there were a lot of people about. When we finally arrived, we parked the car so that we could cut off any escape down the side roads, and went towards the entrance. People were coming out of the disco still fighting, so we picked them up on the road outside. Fortunately reinforcements soon arrived and tempers quickly cooled. After the area was cleared, we found the defendant by the main entrance. Blood was pouring from a bad cut on the back of his head. As it was clear from this that he had been involved, we decided to arrest him.

If you want to sort out all the evidence for yourself, you'll need to look at each account several times and you shouldn't read on until you think you've found the answer.

There are two main differences between what Peter and the bouncer each say. First of all, they disagree about when the incident took place. Peter says that he waited 'for a few minutes' until after he had heard the police sirens before making his way down the steps and towards the doors. The bouncer, on the other hand, claims that it all happened very quickly right at the beginning, and that after he had thrown Peter out, the fighting was still going on inside so he 'went back to help sort out the mess.' They also disagree about how Peter received his injuries. The bouncer says that he was beaten up by other people at the disco because he was running wild. Peter claims that he was hurled backwards against a wall.

Turning to the third account, there's a possible solution. From what we know, there's no reason to disbelieve the policeman's evidence and his report makes two important points. Peter was bleeding from a cut to the back of his head, and they didn't pick him up until quite late, after everybody had been sorted out. Both points suggest that Peter is telling the truth and that the bouncer is trying to protect himself. After all, if the bouncer's story is to be believed, when Peter was put out on to the street, he should have been quite capable of getting away when he heard the police arriving. Furthermore, he'd be very unlikely to get a cut on the *back* of his head by running at a group of people in order to start a fight. If, however, he had been thrown against a wall . . .

We can't tell for certain, but by reading carefully we can at least form an opinion about what is most likely to have happened.

AIMS AND OBJECTIVES

Most of the syllabuses for the GCSE have a section at the beginning which lists the different skills that are needed to do this kind of reading. It's worth spending a bit of time looking at one example of what they say and, if you have worked carefully through the introduction to this section, you might be ready to make some sense of it. This is what the Northern Examining Association consider that you need to be able to do:

The paper will assess the candidates' ability to understand as follows:
(a) **literal: finding the answers to questions when these can be obtained directly by reference to a phrase or sentence in the text;**
(b) **reorganisational: expressing facts, ideas or arguments in their own words;**
(c) **inferential: reading between the lines;**

(d) evaluative: comparing the passage with another, or the passage or part of the passage with their own experience;

(e) appreciation: awareness of language usage, structure, intention and style;

(f) ability to put themselves in the position of characters and appreciate their situation.

The fact that the aims are listed like this doesn't mean that the questions will neatly match each of the points *(a)* to *(f)*. What it does mean is that the examiners will bear these general points in mind as they write their papers. Try to put yourself in their position. Instructions have come through the post from the exam board — copies of the syllabus, past papers, and a deadline that rapidly seems to be getting too close for comfort. A draft paper needs to be written and it's not just a matter of coming up with some good questions, but of finding the right passage in the first place. Possible ideas are picked up, considered and discarded. Eventually a piece of writing is found that seems to be appropriate.

It comes from a book called *I'm the King of the Castle* by Susan Hill, about two ten-year-old boys, Hooper and Kingshaw. Kingshaw's mother is 'housekeeper' to the Hoopers and, when the book opens, she and her son are about to move in. Hooper resents their intrusion into his life.

The crow dived again, and, as it rose, Kingshaw felt the tip of its black wing, beating against his face. He gave a sudden, dry sob. Then, his left foot caught in one of the ruts and he keeled over, going down straight forwards.

He lay with his face in the coarse grass, panting and sobbing by turns, with the sound of his own blood pumping through his ears. He felt the sun on the back of his neck, and his ankle was wrenched. But he would be able to get up. He raised his head, and wiped two fingers across his face. A streak of blood came off, from where a thistle had scratched him. He got unsteadily to his feet, taking in deep, desperate breaths of the close air. He could not see the crow.

But when he began to walk forwards again, it rose up from the grass a little way off, and began to circle and swoop. Kingshaw broke into a run, sobbing and wiping the damp mess of tears and sweat off his face with one hand. There was a blister on his ankle, rubbed raw by the sandal strap. The crow was still quite high, soaring easily, to keep place with him. Now, he had scrambled over the third gate, and he was in the field next to the one that belonged to Warings. He could see the back of the house. He began to run much faster.

This time, he fell and lay completely winded. Through the

runnels of sweat and the sticky tufts of his own hair, he could see a figure, looking down at him from one of the top windows of the house.

Then, there was a single screech, and the terrible beating of wings, and the crow swooped down and landed in the middle of his back.

Kingshaw thought that, in the end, it must have been his screaming that frightened it off, for he dared not move. He lay and closed his eyes and felt the claws of the bird, digging into his skin, through the thin shirt, and began to scream in a queer, gasping sort of way. After a moment or two, the bird rose.

He scrambled up, and ran on, and this time, the crow only hovered above, though not very high up, and still following him, but silently, and no longer attempting to swoop down. Kingshaw felt his legs go weak beneath him, as he climbed the last fence, and stood in the place from which he had started out on his walk, by the edge of the copse. He looked back fearfully. The crow circled a few times, and then dived into the thick foliage of the beech trees.

After a moment, Kingshaw glanced away, turned slowly, and went up between the yew trees and into the house, by the back door.

'You were scared. You were running away.'

'This is my room, you can't come in here just when you want, Hooper.'

'You should lock the door, then, shouldn't you?'

'There isn't any key.'

'Scared of a bird!'

'I was not, then.'

'You were crying, I know, I can tell.'

'Shut up, shut up.'

'It was only a *crow*, a crow isn't anything, haven't you ever seen a crow before? What did you think it would do?'

'It . . .'

'What? What did it do?' Hooper puckered up his face. 'Was it a naughty crow, then, did it frighten Mummy's baby-boy?'

Kingshaw whipped round. Hooper paused. The recollection of Kingshaw's fist on his cheekbone was vivid. He shrugged.

'Why did you go off, anyway? Where did you think you'd get to?'

'Mind your own business. I don't have to tell you anything.'

'Shall I tell you something, Kingshaw?' Hooper came up close to him suddenly, pressing him back against the wall and breathing into his face, 'You're getting a very rude little boy, aren't you,

you're very cocky all of a sudden. Just watch it, that's all.'

Kingshaw bit him hard on the wrist. Hooper let go, backed a step or two, but went on staring at him.

'I'll tell you something, baby-baby, you daren't go into the copse.'

Kingshaw did not reply.

'You went and looked and stopped, because you were a scaredy, it's dark in there.'

'I changed my mind, that's all.'

Hooper straddled a chair beside the bed. 'All right,' he said, in a menacing, amiable voice, 'O.K., go in there, I dare you. And I'll watch. Or into the big wood, even. Yes, you daren't go up into the big wood. If you do, it'll be O.K.'

'What will?'

'Things.'

'I'm not afraid of you, Hooper.'

'Liar.'

'I can go into the wood any time I want.'

'Liar.'

'I don't care if you believe me or not.'

'Oh, yes you do. Liar, liar, liar.'

Silence. Kingshaw bent down and began to fiddle with his sandal strap. He had never been faced with such relentless persecution as this.

'I dare you to go into the copse.'

'Oh, stuff it.'

Hooper stuck his hands up on either side of his head, and waggled his fingers about.

The opening paragraphs in which the attack by the crow is being described need to be read quite carefully, so, if you were an examiner, you might want to start with a question to check whether they've been fully understood. One way of doing this would be to ask for a list of all the injuries that Kingshaw sustains: the wrenched ankle, the bloody face, the blisters and so on. You might decide, on the other hand, that it would be better simply to require a brief account of what happens, in order to see if candidates can identify each stage in the incident. Either way, you'd be testing skills that are labelled 'literal' or 'reorganisational' in the NEA 'aims and objectives'. The answer is there, as long as the right place in the passage can be found.

It's not just because of the opening description, however, that you chose this extract in the first place. Much more menacing, and potentially interesting, than the bird, is the boy – Hooper. Obviously you'll want to ask some questions about him, and his relationship with Kingshaw. This is a bit more difficult, since nowhere does Susan Hill actually *say* what Hooper is like. This is the point, then, at which the candidate will have to 'read between the lines' ('inferential' understanding). It's not very difficult to see that he is determined to persecute Kingshaw in every possible way he can find, but you might want to find a question that would dig a bit deeper than this. He's a bit of a coward, since he backed off when Kingshaw 'bit him hard on the wrist', and paused when Kingshaw 'whipped round'. His behaviour, and the way he talks, also seem very childish. You've got to phrase your question in the right way if you are going to bring out these points. You might ask, for example, if there's anything in the passage to suggest that Hooper is more scared than he appears to be, or you could try a question about the ways in which he taunts Kingshaw and how far he's prepared to go.

Turning to Kingshaw himself, the same kind of question needs to be asked, probably about the way in which he reacts to Hooper and what this tells us about him. It's quite important, for example, to appreciate how well he stands up to the 'relentless persecution', even though he's always on the defensive. A question about whether he is afraid or not might do the trick.

The 'aims and objectives' talk about 'appreciation' so you need to ask something about the writer's techniques. It's very noticeable that Susan Hill hardly comments at all about how the two boys feel during their exchange. It's almost like a play script. You could ask how effective that is, or you could go back to the beginning of the passage and ask for words and phrases to be picked out that are particularly important in making the crow seem so frightening.

Finally, you've got to allow candidates to 'put themselves in the position of characters and appreciate their situation'. There are several ways you could approach this. One obvious solution would be to ask for the passage to be continued. What happens next? Alternatively, you could ask how 'realistic' it is. This would have the advantage of being what the 'aims and objectives' call 'evaluative'.

And that, bar some work on tidying up the wording of the questions, is that. If you can find a suitable passage and try devising questions for yourself, you'll discover that you can often learn more from having to think about what questions to ask, than you do from trying to answer them.

There's one final point that this account of 'aims and objectives' has so far ignored. The National Criteria refer to evaluating 'information in reading material *and in other media*'. That short phrase is very important. 'O' level and CSE rarely ventured beyond an extract of continuous writing, usually taken from a novel. GCSE opens the door to all kinds of material: graphs, pictures, adverts, charts of figures, time-tables, computer printouts, statistics, photos, even video and film. It recognises that a lot of our information comes to us in a much more visually varied form than it used to. Timetables and adverts may always have been the way they are now:

TIMETABLE
CITY COURIER — UNITED SERVICE 40

MONDAY TO FRIDAY 10th June to 6th September 1985 inclusive (not August Bank Holiday Monday—see below for Sunday service)

DURHAM RAILWAY STATION	D.L.I. MUSEUM	NORTH ROAD (JR MILBURGATE)	LEAZES BOWL (COACH PARK)	DURHAM CATHEDRAL	LEAZES BOWL (COACH PARK)	MILBURNGATE (JR NORTH ROAD)	DURHAM RAILWAY STATION	D.L.I. MUSEUM
10.55	→	10.58	11.00	**11.05**	11.10	...	...	...
...	...	...	11.10	**11.15**	11.20	...	...	...
			Then as required (approx. every 10 minutes) until					
...	...	...	1.50	**1.55**	2.00	...	...	...
...	...	...	2.00	**2.05**	2.10	2.12	2.15	2.20*
2.15	2.20*	2.23	2.30	**2.35**	2.40	2.42	2.45	2.50*
2.45	2.50*	2.53	3.00	**3.05**	3.10	...	...	...
...	...	...	3.30	**3.35**	3.40	3.42	3.45	3.50*
3.45	3.50*	3.53	4.00	**4.05**	4.10	...	...	...
...	...	...	4.30	**4.35**	4.40	4.42	4.45	4.50*
4.45	4.50*	4.53	5.00	**5.05**	5.10	...	...	...
...	...	...	5.10	**5.15**	5.20	...	...	...
...	...	...	5.20	**5.25**	5.30	...	...	...
...	...	...	5.30	**5.35**	5.40	5.42	5.45	...
5.45	→	5.48	5.50	**5:55**	6.00	6.02	6.05E	...

NOTES: * - Does not call at D.L.I. Museum on Mondays, when Museum is not open
E - Continues to Elvet Hill Road (arr. 6.10) on request to Driver.

SUNDAYS 9th June to 8th September 1985 inclusive, also Bank Holiday Monday 26th August 1985.

DURHAM RAILWAY STATION	D.L.I. MUSEUM	NORTH ROAD (JR MILBURGATE)	ELVET HILL (ORIENTAL MUSEUM)	LEAZES BOWL (COACH PARK)	DURHAM CATHEDRAL	LEAZES BOWL (COACH PARK)	ELVET HILL (ORIENTAL MUSEUM)	MILBURNGATE (JR NORTH ROAD)	DURHAM RAILWAY STATION	D.L.I. MUSEUM
10.55		10.58	→	11.00	**11.05**	11.10	11.15	...	...	...
...	...	...	11.20	11.30	**11.35**	11.40	...	...	...	...
...	...	...	...	12.00	**12.05**	12.10	12.15	...	...	...
...	...	...	12.20	12.30	**12.35**	12.40	...	...	...	...
...	...	...	...	1.00	**1.05**	1.10	1.15	...	...	...
...	...	...	1.20	1.30	**1.35**	1.40	→	1.42	1.45	1.50
1.45	1.50	1.53	→	2.00	**2.05**	2.10	2.15	...	...	...
...	...	...	2.20	2.30	**2.35**	2.40	→	2.42	2.45	2.50
2.45	2.50	2.53	→	3.00	**3.05**	3.10	3.15	...	...	...
...	...	...	3.20	3.30	**3.35**	3.40	→	3.42	3.45	3.50
3.45	3.50	3.53	→	4.00	**4.05**	4.10	4.15	...	...	...
...	...	...	4.20	4.30	**4.35**	4.40	→	4.42	4.45	4.50
4.45	4.50	4.53	→	5.00	**5.05**	5.10	5.15	...	...	...
...	...	...	5.20	5.30	**5.35**	5.40	→	5.42	5.45	...
5.45	→	5.48	→	5.50	**5.55**	6.00	6.10	...	...	...

NO SERVICE ON SATURDAYS

FARES		
30p SINGLE JOURNEY	50p ALL DAY TICKET	"Explorer" tickets accepted

But even twenty years ago, you would have been unlikely to find a newspaper article that looked like this:

Last in the class

by JUDITH JUDD

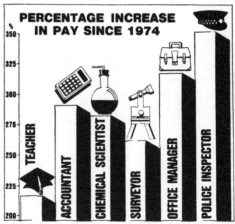

PERCENTAGE INCREASE IN PAY SINCE 1974

TEACHERS' pay has fallen well behind that of comparable professions during the past 10 years.

Government statistics reveal that they now earn less than crane drivers and only fractionally more than deckhands.

In the 10 years from 1974, when the Houghton award was made, teachers' average salaries increased by 217 per cent while those of comparable professions rose by 302 per cent. The professions were agreed to be comparable in last year's pay data working party of teachers and employers.

Since Houghton, only finance clerks and nurses — neither a graduate profession—have had lower ⌐ Poli-

may have been promoted to inspector on £12,242, but a teacher with six years experience can expect £7,734.

Journalists' average weekly pay in April, 1984 was £259.30, according to the Department of Employment's new earnings survey from which the other figures are also taken (Figures for Apr¹¹ ¹⁰⁰ ⁻⁻⁻⁻vailable).

below the rises retail price index.

Local authorities which employ teachers accept that they have fallen behind, but argue that they cannot afford more than the 4 per cent offered in response to the claim for £1,200 extra.

The Government has said that there will be no more money unle⁻⁻ ⁻ ⁻st of

or had an illustration attached to it like this:

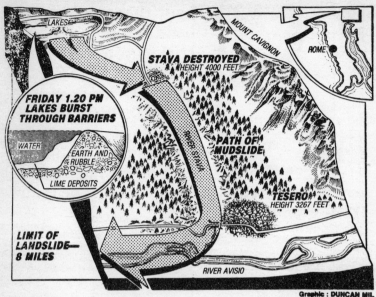

Graphic : DUNCAN MIL

The deadly river of mud released when the dam was breached.

Diagrams like these need to be 'read' just as carefully as normal text. In the graph about teachers' pay, for example, you have to be able to work out how it's been designed. The same figures could have been presented in a different way. In the diagram, the vertical axis, which measures the percentage pay increase since 1974, starts at 200%. If it started at 0%, the diagram would look like this:

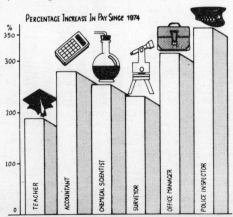

Teachers have still fallen behind over the last ten years, but it doesn't look quite so dramatic, does it? The map showing the path of the dam-burst in Italy also presents problems. You need to be able to understand, for example, the convention whereby, in the top right-hand corner, the artist is able to show where in Italy the disaster occurred. You can be fairly sure that the GCSE will be expecting you to cope with this kind of material as well as more conventional print.

STYLE OF QUESTIONS

If you have been following the argument closely, you'll have spotted that a lot of new demands are being made. There'll still be 'comprehensions' of course, but there'll be a lot more besides, and it won't all fit into one short exam paper as in the past. You may find, then, that even if there's one part of the syllabus dealing exclusively with reading, that won't be the end of it. It might be necessary, for example, to include a piece of work in your coursework folder that could be described as 'writing based on varied reading to demonstrate close reading and informed response' (LEAG), or 'work which shows evidence of the candidate's ability to understand what he or she has read' (NEA). Every syllabus will be different. Nonetheless, it is possible to give a general idea of what to expect, since the exam boards will still be basing much of what they do on past experience. Four examples are provided here. They've been arranged in a logical fashion, starting with tests that require you to do little or no writing, in which answers are either right or wrong, and ending with assignments that allow greater freedom for you to interpret and use what you are reading.

Multiple choice tests

Most candidates will *not* be faced with multiple choice or 'objective' tests, at least in English or English Literature. However, they are allowed by the National Criteria as long as they don't count for more than 20% of the marks available. During the planning stages for specimen papers, the Southern Examining Group produced a sample syllabus which included a multiple choice test; they eventually withdrew it, but it does serve as an example of what objective tests in English are like.

Peter Scott traps the world's top killer

When Peter Scott was two, his father, Scott of the Antarctic, died on the way back from the South Pole. Among his last letters, found with his body, was one containing this plea: 'Make the boy interested in natural history if you can; it is better than games . . .'

5　Peter grew up to be perhaps the world's most famous painter of
birds. He went on to devote much of his life to the study and pre-
servation of wild life of all kinds. I went to see him at Slimbridge,
that isolated settlement by the mudflats and marshes of the Severn
Estuary where, just after the war, he founded the Wildfowl Trust.

10　Here he lives as its honorary director, and, when he has time, paints
for a living. As we talked, he worked on a canvas in his studio, the
window of which is only a few feet from the Trust's magnificent
Swan Lake.

'I love painting,' he said. 'I love every minute of it. But even if I
15　didn't, I would have to go on doing it. We're always overdrawn at
the bank, and this is the only way I can sort them out.'

You will never see an art gallery with a Peter Scott exhibition. Just
about everything he paints is commissioned before he does it —
long before. He showed me the 'outstanding' list; it ranged from
20　a famous actor to the British Museum.

Some specified the exact size required, some the subject. One even
mentioned the date by which the work was wanted, but it had
gone by. If he spent more time at the easel, he could probably
become the richest artist in the country. But a good deal of his
25　extraordinary energy goes to saving animals from extinction as a
result of human ignorance, apathy and brutality.

At least the wild birds have hope, at Slimbridge. Within the 50 fox-
proof acres there is a resident population of 3000. They include
scores of Hawaiian geese, chattering noisily round your ankles.
30　Twenty-one years ago there were only 40 in the world. Three were
sent to the safety of Slimbridge, and eventually Mr Scott was able
to send 90 back to Hawaii.

Beyond the Trust fence, on the quiet miles of the river estuary, as
many as 5000 wild ducks and 8000 wild geese may spend the
35　winter, waiting for their Arctic homes in Northern Europe to thaw.
Hundreds of Bewick swans swim up to Mr Scott's windows, each
individual recognisable by the different pattern of yellow on its
black beak.

Before you can see any of the wildfowl at Slimbridge (closed only
40　on Sunday mornings) you have to pass along a corridor, one wall of
which carries a bold and simple explanation of why Peter Scott
cares about wild life. 'Look into the frame below,' a notice reads,
'and you will see a specimen of the most dangerous and destructive
animal the world has ever known.'

45　You look, of course, and what you see is a mirror.

1 In the first paragraph we learn that Peter Scott was

   A orphaned at an early age

   B the son of a famous father

   C a credit to his parents

   D rather a weak child for his age

2 The letter referred to in the first paragraph was most probably written to Peter Scott's

   A wife

   B headmaster

   C guardian

   D mother

3 According to the passage, his father hoped that Peter Scott would

   A never play football or cricket

   B learn about the lives of animals

   C read plenty of history books

   D study the wonders of nature

4 The line of dots after 'games' in line 4 indicates that

   A this is an extract from a longer passage

   B the writer died before he could finish the sentence

   C the next sentence was written on another topic

   D at this point the paragraph ended

5 The settlement in line 8 is described as 'isolated' because it is

   A surrounded by the waters of a lake

   B situated in the middle of a swamp

   C remote from towns and villages

   D close to the mudflats of an estuary

6 An 'honorary' director (line 10) is one who

   A is not paid for his services

   B is admired by everybody

   C does not do any actual work

   D has been decorated by the Queen

7 Peter Scott told the interviewer that he has to continue painting because

   A it is such an enthralling activity

   B they are constantly in debt

   C there is a long waiting list for his work

   D it is his only source of income

8   The list in line 19 is referred to as 'outstanding' because it contains

   A   names of important people and organisations
   B   very detailed descriptions of the paintings
   C   works that would one day be famous
   D   orders that have not yet been fulfilled

The first thing to notice about these questions is that they are cunningly designed to catch you out. The wrong answers have been just as carefully written as the right ones, so that you have to read very closely in order to ensure that you don't make a mistake. Look at the first question again:

1   **In the first paragraph we learn that Peter Scott was**
A   **orphaned at an early age**

Don't be fooled. Because the passage mentions his father's death and doesn't refer to his mother at all, you could jump to the conclusion that Peter Scott was orphaned. Or, in the heat of the moment, and relieved at having found a phrase ('his father . . . died') that seems to give you the answer, you might forget that 'orphaned' means losing *both* parents. You'll only get your marks, however, by avoiding both of these traps and moving on to consider the next choice.

B   **the son of a famous father**

If you've heard of 'Scott of the Antarctic' you may agree that we do, indeed, learn that Peter Scott was the son of a famous father. But if you are wise, you'll also be cautious. Knowing how these questions work, you'll be wondering what the catch is, and you might be suspicious of this choice since the information is only supplied incidentally – it's not the main point of the sentence. Can it really be what they are getting at?

C   **a credit to his parents**

Another tricky choice. If you have read the whole passage with its account of everything that Peter Scott has achieved, you might well describe him as 'a credit to his parents'. The examiner knows this and the point behind the question is to see whether, in your answer, you can stick closely to what the passage actually says in black and white. It's another choice that is deliberately misleading.

D   **rather a weak child for his age**

This choice works in much the same way as the last one, though you shouldn't have so much difficulty in seeing through it. There's nothing to suggest that he was weak for his age, but, on a hasty reading, you might assume that this is why he shouldn't be encouraged to play 'games'. What you should have realised is that his father's reason for making this 'plea' stems from the belief that an interest in 'natural history' would be more worthwhile.

If that is what goes through your mind as you examine each choice, you might still be uncertain, even when you've finished. All the indications may point to **B** as the correct answer, but can you be sure?

In the end, everything will rest upon how well you can read the examiner's mind. If you understand how multiple choice tests work, then you'll be able to spot the right choice. Consider what you've learnt from looking at the four choices in question **1**:

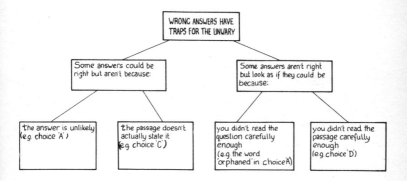

By now, you should be fairly confident. It isn't that choice **B** – the son of a famous father – is obviously right, but rather that it isn't obviously wrong. There are no hidden surprises, no traps, no red herrings in it. And if you chose **B**, you would have chosen correctly.

We supply the rest of the answers. Before looking at them, however, take one or two of the questions and see if you can identify which kinds of traps are being set for you.

There are a number of reasons why multiple choice tests are unlikely to be used by many of the examining boards for the GCSE. Some people say that they are not appropriate for testing English, where a lot of answers are neither definitely right nor definitely wrong. Some people object to the way in which you have to play guessing games with the examiners, and they don't like tests which deliberately offer you wrong answers. Others say that multiple choice tests are very artificial exercises, unlike anything that you'll have to do when you've left school. For all these reasons, a much more common approach is likely to be the use of the short-answer test.

Answers to multiple choice: 1 – B, 2 – D, 3 – B, 4 – A, 5 – C, 6 – A, 7 – B, 8 – D.

The short-answer test

For years, this has been the accepted way of assessing how well you can read, and it's not going to disappear with the introduction of GCSE. At a quick glance most short-answer comprehensions look the same, but different examining boards do have different traditions, and some are better than others. This sample comes from one of the GCSE pilot schemes.

SECTION A

This is the story of a coloured boy's attempt to join a library used only by white people in the southern states of the USA. Read it. Then answer the questions on it.

He didn't know how far he walked before he saw a stone building sitting on a hill. That was probably the library. It looked just like the one in Kansas City, grey and ugly. He walked quickly up the steps, studiously ignoring the white faces he passed. He went into
5 the building as if he'd been there many times and walked to the desk. When the old white woman looked up, her jaw dropped. 'What you want?' she said sharply.

'I'd like to apply for a library card,' he said firmly.

'You can't come to this library,' she said nervously.

10 Allen could feel his heart pounding as he noticed the white people in the library gathering a short distance away. He didn't know what to do, but he knew he couldn't walk out of that that library past all those white faces. He couldn't let them run him away. 'Why not?' he said calmly.

15 'You just can't,' the old woman said, more agitated. She had lowered her head and was busy stamping some cards on her desk.

'I would take proper care of the books.' He spoke distinctly and evenly, betraying no emotion and being very careful not to sound coloured, like his father. And though he was angry, his voice was as
20 pleasant as if he were talking about the weather.

'This is the white library,' the old woman blurted out. 'You people have your own library.'

Allen hadn't known there was a coloured library, but it didn't matter.

25 'But one does not have the wide choice of books there that are available here. And I think it's the duty of all Americans to be as fully educated as they can be. Don't you agree?' He almost burst out laughing and wished his father was there to see him.

The old librarian turned a deep red and refused to answer. When

30 Allen realised that she was going to ignore him, he became frightened. He couldn't let her win. He simply couldn't.

'Is there a law against my availing myself of these facilities?'

'Yes,' the woman snapped.

'Might I see it please? I'm not familiar with it.'

35 'Where you from, boy?' the woman asked evenly, looking at him through narrowed eyes. 'You don't talk like you from Nashville.'

'No, I'm not. I've just moved to the city from Pine Bluff, Arkansas.' And it wasn't a total lie. He had been in Pine Bluff for a week before they came to Nashville. It was obvious, however, that he
40 wasn't going to get a library card. He could sense that a crowd had gathered, and he knew that if he continued to press her something might happen. He didn't know what — she might call the police. But he had to have a library card.

Just then a young white woman came out of a back office. 'Um—
45 Oh,' he thought. The old woman had probably pushed a buzzer under her desk, or somebody went and got this younger one and she was coming out to tell him to leave before she called the cops.

'May I help you?' the woman said pleasantly.

'Yes, I would like to apply for a library card and this woman told
50 me I can't have one. I don't understand why. All I want to do is read.'

'What are you interested in?' the young woman continued.

'Oh,' Allen began eagerly, 'I'd like to see if there's a biography of Winslow Homer. He's one of my favourite painters. And also I'd
55 like the Thayer two-volume biography of Beethoven.' He was sincere, but he was also trying to impress her. She probably thought he was going to list some novels or murder mysteries.

'Well, Mrs Helms,' the younger woman said, 'since I know those books wouldn't be available at the coloured library, I don't think
60 we'd be breaking any rules if we let this young man have a card.'

Allen allowed himself to get happy, but the woman had called him 'young man' and not 'boy' and that made him a little wary. No white woman called a Negro anything but 'boy'.

The old librarian was obviously furious, but she only spluttered,
65 'Whatever you say, Mrs MacIntosh.'

Allen was surprised. The younger librarian was probably in charge of the whole library. The other one probably wasn't even a librarian, but just some ol' white woman who sat there and looked so unpleasant she made people want to read books so they'd forget
70 about her.

'Would you come with me, please?' the younger woman said.

Allen wanted to turn and stick out his tongue at all the white people standing around, but just as he had shown no expression the day he scored twelve points at basket-ball his face was impas-
75 sive now. He walked into the woman's office and she handed him a card to fill out.

'These people are funny, aren't they?' she said.

'I beg your pardon?' he replied cautiously.

'I mean their silly rules. They think the library will fall if coloured
80 people start using it.'

He didn't say anything, knowing that it was particularly unwise to get into conversations with white people when they were talking against other white people. He filled out the card quickly and handed it back to her.

85 'You're only fourteen?'

'That's right,' he said pleasantly.

'Aren't you mighty young to be reading such difficult books?'

'I don't think so.'

'Well, we'll have to have your mother or father's permission. Take
90 this card home,' she said, handing him another card, 'and have your mother sign it and bring it back as soon as you can. In the meantime I'll make out a temporary card for you so you can take some books out today. When you bring this other card back with your mother's signature, we'll give you a permanent card.' She sat down
95 at the typewriter and quickly typed out the temporary card. 'I don't know if you know it, but you're the first coloured person to use this library.'

'I didn't know.'

'There shouldn't be any trouble though. But this could cost me my
100 job.'

He felt a little guilty.

'I don't think so,' she continued. 'They had me come down here from Ohio to take this job, and I don't think they'll fire me just yet.'

105 So that was it. She was from the North. He wanted to apologise to her for maybe causing her trouble, but he didn't. He hadn't done anything wrong.

'Let me show you around the library so you'll know where the various books are.'

110 'Oh, that's all right,' he said quickly. 'I can find everything on my own.'

'It's no trouble.'

Allen glumly followed her out the door. The last thing he wanted was to be shown around the library by a young white woman. It
115 was bad enough that he was there. But the librarian didn't notice his discomfort. The library lobby was empty now, Allen was glad to note, and the old woman didn't even look up as they passed her desk. He hardly listened as the librarian took him through the stacks, showing him what books were shelved where. 'Here's that Thayer
120 you wanted,' she said, bending down and taking two thick volumes from a lower shelf and handing them to him.

'Thank you.' He held the books in his hands for a moment. They were dusty, but he lifted them to his nose and inhaled. There was nothing like the smell of old books.

125 'Do you like Beethoven?' she asked.

'Yes, but I haven't heard that much,' he admitted. He didn't tell her that his real interest in Beethoven was in the fact that he'd read somewhere that when the composer was a boy he was so dark he was called 'Spangy'. Allen wanted to find out if Beethoven was
130 really coloured.

'Ah, and here's a book on Winslow Homer.'

He took the book from her, anxious to get out of there and run home to report his adventure.

QUESTIONS:

A1 You must study carefully lines 1 — 43 in order to answer the two parts of this question.

(a) Tell the story of what happens in these lines in about 100 words. Avoid using words spoken by the characters.

(6 marks)

(b) Point out the differences in the ways Allen and 'the old white woman' speak and behave. Show why Allen is able to embarrass and annoy her.

(8 marks)

A2 Study lines 44 — 133. Describe the character of Mrs MacIntosh, 'the young white woman'. For each point you make about her character give an example from these lines to support it.

(6 marks)

A3 What reasons do you think Allen had for going into the library other than to obtain a library card?

(4 marks)

A4 What reasons do you think Mrs MacIntosh had for letting Allen have a library card other than kindness?

(4 marks)

A5 Read again lines 48 — 84. Point out how in these lines Allen shows intelligence and quick thinking.

(8 marks)

A6 From this story what impressions have you formed about the type of society in which these people live?

(4 marks)

This passage and the questions with it will have been set by a 'Chief Examiner'. To mark the answers, the board employs a team of assistant examiners, all of whom must apply exactly the same standards. So the Chief Examiner usually produces a 'marking scheme'. What follows is an informed guess about the guidance that might have been provided in the marking scheme for this paper. For the first two questions, instructions have been provided in full. In A2-A6 gaps have been left in order to encourage you to think about this kind of test from an examiner's point of view. See if you can fill them in.

A1 You must study carefully lines 1 - 43 in order to answer the two parts of this question.

(a) Tell the story of what happens in these lines in about 100 words. Avoid using words spoken by the characters. *(6 marks)*

Do not award any marks for references to events beyond line 43. Candidates must have included reference to:

Allen going into the 'whites only' library

Being informed that he isn't allowed in

Pretending he doesn't know why he is being refused entry and saying he will look after the books

Saying that the coloured library is inadequate

Asking to see the law which forbids him to use the library

Being asked where he comes from and replying

The crowd gathering

Please give one mark for each point (6 maximum), don't reward trivial details not of central relevance to the story (e.g. description of the outside of the library/librarian stamping books when Allen enters). This is a fairly simple question in which candidates have only to identify the outline of the story. Many will get full marks.

(b) **Point out the differences in the ways Allen and the 'old white woman' speak and behave. Show why Allen is able to embarrass and annoy her.** *(8 marks)*

This question is more difficult and some candidates will be unable to read closely enough to see the more subtle differences between them.

Use 4 marks, but no more, for candidates able to show that:
Allen remains calm ('firmly'/'calmly'/'betraying no emotion'), whereas the white woman becomes flustered ('nervously'/'more agitated'/'snapped')

Allen speaks in Standard English ('You don't talk like you from Nashville'/ e.g. 'Is there a law against availing myself of these facilities'), whereas the white woman speaks in a local dialect (e.g. 'Where you from, boy?')

Allen is polite ('evenly'/'his voice was pleasant') whereas the white woman is rude ('sharply'/e.g. 'You just can't')

Other examples are acceptable as long as they are clearly related to the point being made.

In the second part of the question, candidates should be able to spot that the old white woman is embarrassed and annoyed because:

Allen is determined
he has a well thought out argument to which she has no answer
he has the nerve to ask to see the law
he behaves very politely

Extra marks are available to candidates who are able to point out that Allen's educated and civilised behaviour is contrasted with the old white woman's rudeness in order to make a mockery of the laws against blacks.

A2 **Study lines 44 – 133. Describe the character of Mrs MacIntosh, 'the young white woman'. For each point you make about her character give an example from these lines to support it.** *(6 marks)*

Any points about her character are acceptable as long as they are supported with examples — one mark for each point, please, and one for each example. Candidates might include the following:

character	example
polite	she calls Allen 'young man' not 'boy'
courageous	she says 'this could cost me my job'
. . . .	. . . .
. . . .	. . . .

A3 What reasons do you think Allen had for going to the library other than to obtain a library card? *(4 marks)*

This question tests whether candidates can understand things that are *implied* in the passage. Acceptable answers will include the following unstated reasons (2 marks for each):

In order to challenge the laws against blacks using the 'whites only' library.

In order to borrow particular books that he wanted to read and couldn't obtain elsewhere.

A4 What reasons do you think Mrs MacIntosh had for letting Allen have a library card other than kindness? *(4 marks)*

This tests the same skills as the last question. Acceptable answers will make reference to the following (2 marks for each):

. . . .

. . . .

A5 Read again lines 48 – 84. Point out how in these lines Allen shows intelligence and quick thinking. *(8 marks)*

Candidates need to be able to judge how well Allen handles the situation. Most answers will include the following, but other reasonable points will be accepted (2 marks for a well made point):

1 He asks for difficult and serious books.

2 . . . .

3 . . . .

4 . . . .

A6 From this story what impressions have you formed about the type of society in which these people live? *(4 marks)*

Answers to this question must be marked by impression. It asks candidates to tackle the passage as a whole. Acceptable answers will include observations about the racial laws in force in America, good answers will also comment on . . . .

. . . .

. . . .

The 'response'

All the different lists of aims and objectives being produced for the various GCSE syllabuses have one thing in common. When they are dealing with reading, they use the word 'response'. The reason for this is that the National Criteria, the guidelines for the exam, say that 'opportunities must be provided for pupils . . . to *respond* in a variety of ways to what is read'. As a result, all the exam boards are searching for something that doesn't look like a 'comprehension' test, either multiple choice or short-answer, but which nonetheless provides a way of judging how well you can read.

A couple of examples will give you an idea of how the problem is being solved. The multiple choice paper about Sir Peter Scott and the Wildfowl Trust had two parts to it. The second part included the following question:

Using your own words as far as possible, write a publicity leaflet about the Wildfowl Trust at Slimbridge. Your leaflet should include an account of the foundation of the Trust, and references to Peter Scott's position there as well as to his painting, the geographical situation, purpose and achievements of Slimbridge and the continuing threat to wild life.

In the Midland Examining Group specimen examination, there's a similar question. After a lengthy newspaper article describing how old people can be tricked into letting people into their houses, candidates are asked to:

Write a leaflet designed to inform old people about how to protect themselves from unwelcome visitors. Your advice should be helpful but not alarming and may include both information from the report and ideas of your own.

As far as the assessment of reading is concerned, it's this part of the syllabus that is going to be most unfamiliar. Both of those questions depend upon your having read and understood the passage (you can't do them well if you haven't), but they're very different from the multiple choice or short-answer tests. Instead of answering a direct question about a piece of writing, you will be expected to *use* the information you have gathered to do something which *shows* that you have understood. Put more simply, the question will give you information in one form and ask you to change it into another. A diagram might help to show you what is involved.

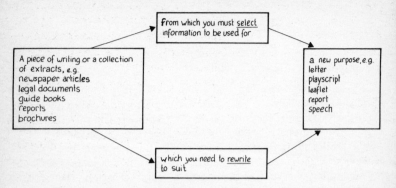

The two key words which describe what you need to be able to do are 'select' and 'rewrite'. In practice, this is quite a skilled job but, unlike the multiple choice test, it does at least match the kind of reading and writing tasks that will be expected of you when you leave school.

Sometimes, you will find that you have to come to terms with quite a lot of reading material. The complete version of the assignment that follows (from SEG) runs to four sides and includes a page from a catalogue with pictures of different kinds of sports shoes, a mail order form and the following information about your legal rights as a consumer:

(i) Defective goods

Knowing what a 'defect' is and how to claim for it.

A trader who sells a customer goods that are defective, faulty or otherwise unsatisfactory must pay compensation or refund his money in full if:

* **The goods do not match the seller's description.**

* **The goods are not fit for their common, everyday use.**

* **The goods do not do the specific job for which they were bought.**

(ii) When goods are not as described

Goods are defective if they do not match any oral or written description attached to them at the time of the sale.

If, for example, a ring described as 'solid gold' turns out to be gold-plated, the trader has broken a term of his contract with the buyer and is liable to compensate him.

Similarly, if goods are chosen from a sample — for example, wallpaper or carpets — they must conform to the sample in style, colour and texture.

(iii) Goods not fit for their purposes

Any article you buy must be reasonably fit for the purpose to which people usually put goods of that kind – what the law calls of 'merchantable quality'. In addition, if you have told the trader you want an article for a specific purpose or job, it must be suitable for that.

Merchantable quality. Goods need not be perfect or without blemish to satisfy the 'merchantable quality' condition. The test is whether a reasonable person, knowing their condition and their normal, everyday usage, would still buy them for the price.

But a trader cannot escape his obligation to provide goods of merchantable quality simply by labelling them 'seconds' or 'defective'. They must still be as suitable for their general purpose as it is reasonable to expect 'seconds' to be.

If they turn out to be less than suitable, the trader can escape liability to pay compensation only if he pointed out, at the time of sale, the particular defect that later caused complaint.

Specific suitability. An article can be of merchantable quality but be unfit for the specific purpose you have in mind. A length of material, for example, could be used to make curtains or a skirt or cover an armchair.

If you wanted to use it for upholstery, and it proved unsatisfactory for that, though it was suitable for its other purposes, you would have no claim against the trader unless you had told him why you wanted the material.

When a trader need not pay

When you take goods back to a shop on the ground that they are not of merchantable quality, your claim will not succeed if:

* The trader pointed out the specific defect before you bought the article.

* You noticed the defect or examined the article closely enough to have been able to notice it.

* You may claim that the goods are unfit for a specific purpose. The trader need not pay compensation if he can show that you ignored his recommendations and bought something he had advised against.

Asking for your money back

Assuming the trader has broken one of the conditions of sale, you must still act quickly if you want to return defective goods and get your money back.

* Do not keep the goods longer than is reasonably necessary to discover the defect.

* Do not tell the trader that the goods are satisfactory and that you are keeping them.
* Do not consume any of the goods.

You are entitled to some delay, however, if the goods you bought — tinned food, for example — could not be fully examined until you came to use them.

Claiming compensation

Even if you do not reject the goods promptly, you have the right to claim compensation for up to 6 years from the date of purchase — unless you are claiming compensation for personal injury caused by the goods. In that case the limit is 3 years.

There are two main types of compensation, both of which are intended to put a claimant in the same financial position as if the goods had not been defective:

* If the defect can be put right by having the goods repaired, the trader must pay any reasonable repair cost.

    You are not obliged to give the goods back to the retailer or manufacturer. You can have the repair work done anywhere.
* If the goods cannot be repaired, you can ask the trader to pay you compensation in the form of a partial refund. In effect, he pays you the difference between the true value of the defective goods and the amount you originally gave him for them.

One of the three questions set is this:

Shortly after the shoes arrived, the soles came away from the uppers. You wrote to ask for your money back and the firm refused claiming that you had misused the shoes. You now decide to take further action. Using the relevant information from pages 62 to 64, write a formal letter to the Managing Director of the firm, justifying your claim for compensation.

Remember the diagram? Your first task is to *select* the relevant details. You can ignore, for example, the whole of section (ii) 'When goods are not as described', but you have to read (iii) 'Goods not fit for their purposes' much more carefully. If the only thing you did with the shoes was to put them on your feet, then it's the first sentence of section (iii) that is really important.

Having sorted out the information that you need, you then have to *rewrite* it for the letter. Otherwise, it would read very oddly:

Any article you buy must be reasonably fit for the purpose to which people usually put goods of that kind —what the law calls of 'merchantable quality'.

Before putting pen to paper, however, you must take into account the situation, since it will help you decide both *what* to say and *how* to say it. The question points out that this is your second letter of complaint. That suggests that whilst you should refrain from becoming abusive and threatening, you can word things a bit more strongly than you might otherwise have done. With all this in mind, you can finally make a start. When you come to the appropriate part of the letter, you might want to rephrase the legal reference something like this:

> Since buying the shoes from your firm, I have used them only in ways for which they are designed, as normal footwear and as running shoes.

The thing that you mustn't forget in this kind of work is that it will be marked, at least in part, as a test of how well you can *read*. Your final product must show that you have taken into account all the information that you were given to start with.

This section has been about 'response' and assumed that, as with the Southern Examining Group from which the last example was taken, you will be set an examination paper. However, it's quite possible that assignments like this will also be included in your coursework.

Coursework

One thing that a book like this can't do is to guess at the contents of your coursework folder. Too much depends upon what you and your teacher decide to put there. However, there's one problem that is so easily solved by coursework and so difficult to solve any other way that it seems almost certain to become a requirement.

The National Criteria say quite clearly that syllabuses must include 'sustained reading of various kinds, including the reading of whole works of literature rather than extracts only'. What's more, they 'must not require the study of prescribed texts'. It's hard to see how that could be tested in an examination.

This doesn't mean that GCSE 'English' is 'English Literature' in disguise. The examiners want to encourage 'sustained reading' because they would feel uneasy if, at the end of the course, you were left with the impression that all people ever did with books was read short extracts from them. After all, one of the pleasures of reading comes from the way in which a story unfolds over a hundred or two hundred pages.

You will probably find in 'English' that you have much more freedom of choice about how you respond to the books you have read than in 'English Literature', and that you won't be restricted to the traditional type of literary essay which expects you to discuss character, plot and themes. A lot of what you do will be suggested by the books themselves, but here are some general ideas that might be applied to almost any book you read:

1 continue the story using the same characters;
2 from the point of view of one of the characters in the book, write a letter, or a diary entry, describing some of the events which you have experienced;
3 write a conversation between two of the characters about one of the other people in the book;
4 describe how you would turn the book into a film, or a television serial, or a ballad, or a cartoon;
5 write articles for a newspaper or a magazine that could have been prompted by some of the events in the book;
6 give a personal account of the book and what it means to you;
7 review the book for somebody else, deciding whether you would recommend that they should read it.

This list is not complete, but it should let you know what to expect in this part of the exam, and it might give you some ideas if you do have the freedom of action that coursework can allow.

Regardless of the particular way in which you are assessed, there are some things that you always need to bear in mind when confronted with a piece of writing.

GOOD READING HABITS

Understanding a text

Words aren't simple. Dictionaries, which attempt to provide accurate definitions, can never paint a complete picture, because the meaning of a word is affected by the way in which it is used. As you turn the pages of a book you are probably no more than half aware of what each word is doing, but an effect is created nonetheless. Part of the business of understanding a text lies in the skill with which you are able to see how this is happening.

This point becomes clearer if you try reading a text that has been tampered with in order to see whether you can reassemble it. A number of words have been removed from the following passage. There should be sufficient clues in what remains, however, to allow you to complete the gaps in a way that makes sense. It doesn't matter how the original version was written, what is important is whether you can provide alternatives that fit. You'll only be able to do it well if you examine the passage very carefully indeed, and it works better if you can discuss your choices with a friend.

Smoking

If you are a smoker $-----$ to give it up, you are a $-----$ of a large club. More than half the $-----$ population of Britain smokes, $-----$ half of them $-----$ to give it up. $-----$, more than four-fifths of all smokers $-----$ to their last breath, even though they know how $-----$ that last breath is $-----$ to be.

The high $----- -----$ in attempts to give up smoking is because, for $-----$ smokers, it is more an $-----$, than a $-----$ pleasure. Many smokers might deny this. But smoking is $-----$ not a pleasure in the $-----$ that eating fish and chips or $-----$ to Mozart or riding on roundabouts can be $-----$. People are not normally bad-tempered $-----$ they have not $-----$ their first Mozart record or roundabout ride of the day. $-----$ there cannot be many people so $-----$ by passion for fish and chips that they $- - - - -$ a deserted town late at night looking for a $- - - - -$ of supply.

Breaking an addiction is different from giving up a pleasure. The $-----$ for breaking this particular addiction are numerous. For whatever the $-----$ about the extent of smoking's contribution to various illnesses, there is no medical expert who $-----$ that it does you good.

This next exercise works on the same principle. It is based on the idea that, as long as you are reading carefully, you can pick up unconscious clues about what is going to happen next. You'll need to use a piece of paper to cover up all but the first section of the story. Start by reading this, trying to work out what is going on. Then predict, on the basis of any evidence you can find in the language that is used, what is going to be contained in the next section. Keep going in this way down the pages, referring back to work out why your guesses were right or wrong.

1   When she at last got back to the farm-yard, ready to shut up the hens for the night, she was suddenly aware of the figure of a man disappearing towards the woods, behind a stack of hay.

    She was quick to raise the shot-gun and started running. She caught another glimpse of the man climbing a stile. Another two-legged fox, she told herself, and shouted:

    'Here, you there! What's the big idea, traipsing all over other people's property?'

2   He stopped and turned. She was still running as he turned, but suddenly she stopped too. Gun still raised, she found herself face to face with a thin, boyish young soldier.

'What the blazes do you think you're up to? Nicking something I shouldn't wonder.'

3   He started to mutter something about losing his way. His intensely blue eyes seemed scared. He was carrying his forage cap tucked in a shoulder strap. His fair light hair shone almost white.

'What are you doing up here, on other folk's land? Don't they give you nothing to do in the army nowadays?'

'Walking. That's all. Just walking.'

'Funny you should walk into my place.'

'Just got lost. A bit lost — '

'Well, you can get lost again. I don't want nobody trespassing and traipsing about up here.'

'I thought there might be a way, foot-path or something, down the hill — '

'That's the only path there is. Through the wood. That way.'

She lifted the gun again, pointing it straight at him. This time he seemed more offended than scared.

4   'Don't you know you should never point a gun at anybody?'

'Sorry,' she said and surprised herself in saying it. 'Sorry.'

She lowered the gun. There was a good ten seconds of silence between them. Once or twice he ran his fingers through his hair. As it moved it seemed whiter than ever, almost luminous against the darkening background of tree shadow.

5   'You see, I — '

She found herself starting a confused explanation. Well, she was all alone up here. You never knew about people. She was always afraid there might be somebody —

'You're not afraid of me, are you?'

'I didn't say that. I didn't mean that.'

'No need to be afraid of me. I'm nobody. My name's Barton.'

All the time the light blue, boyish eyes seemed to indicate more and more that it was he, not she, who was afraid.

6   'I had things nicked before now. Hens. I know what soldiers are. Scrounging. On the scrounge all the time. I know. My husband's a soldier.'

'Nightingale,' he suddenly said. 'Nightingale.'

He lifted his head listening. She could hear the nightingale too, somewhere far up, hidden in the beech branches. For fully half a minute they listened to it together until at last he said:

'Marvellous, that. Marvellous.'

7   She said nothing.

'Haven't heard one for a long time,' he said. 'Used to have them where I come from but they've all gone now.'

'Where's that, you come from?'

'Hampshire. Like this. Not far from the sea.'

Up in the beech trees the nightingale held to one long pure sustained high note. The soldier drew in his breath and held it too.

'Nothing like it,' he said. 'They say some birds can imitate it but — '

'They sing so much I hardly seem to notice it.'

'They sing all day sometimes. Do they up here?'

'Like I said, I hardly seem to notice it.'

Now there was another long pause between them and again they listened to the nightingale. This time she broke the silence by saying:

8   'Sorry about the gun. I ought to have known better than that.'

Whether this embarrassed him or not she never knew. He simply said:

'We've got a place like this down in Hampshire.' He looked about him for a second or two. 'Bit bigger though. My Dad and me used to run it together. He's on his own now. Hard graft for him, all on his own.'

'I should know.'

Sharply his brilliant blue eyes turned to the direction of the high beech tops and the nightingale.

'I could listen all night to that. I could just stop and listen all night.'

To her own great surprise she actually found herself laughing.

'I can see you turning up at the guard room,' she said, 'with a nightingale for an excuse. What would they say to that?'

9   'Put me on a charge,' he said and laughed too. 'Worth it though.'

There was now yet another silence and it was she who broke it again.

10   'I was just going to have my supper.' She again felt con-science-stricken by her stupidity about the gun. She wanted, somehow, to make up for it. 'You could come in for a few minutes if you'd like. I'm sorry I spoke like that.'

If you've done both of these exercises properly, you should have found that you were putting every word under the microscope. It's only if you get into the habit of doing this, that you'll start to make real improvements.

SUMMARY

1 Words mean more than their dictionary definitions.
2 You need to read and re-read carefully if you are going to understand the full meaning of a word.

Study techniques

Don't be mesmerised into thinking that you should always start reading a book at the beginning and carry on doggedly to the final page. Even with a novel, which probably does need to be read in that way, you will often find that there's an editor's introduction which is printed first but can be read last. Reference books, like the cookery book mentioned earlier, positively demand *not* to be read page by page. For the most part, they will have a *list of contents* at the front and an *index* at the back to help you find your way around in the way that best suits your purpose.

In order to use a list of chapter headings properly, however, you need to be able to *skim* and *scan*. A *skim* read is a quick read through the text, in order to get a general impression of what it is about. That allows you to identify the particular passage you want, which can then be *scanned*.

It's this second kind of reading that will mostly be tested by the GCSE, and it's no use imagining that you can hope for success just by running your eyes conscientiously down the page. Good reading of this kind should be an *active* process. Don't be frightened of taking a pen or pencil to the passage, though you should be sensitive about this if you are working directly from a book, and always make sure you have some scrap paper to make notes on. The best way of reading *this* book, for example, if it's your own book, would be for you to use a marker pen to highlight important points, or to scribble comments in the margin, or to do some of the exercises that are suggested.

Imagine, for example, that you were doing some work on 'Law and order'. It's leading up to an assignment for your coursework folder, but as preparation for the written work the teacher has asked everybody in the class to do some reading. A number of books have been provided, and you've chosen to use one from the Longman Social Studies series called *Crime and Punishment.* You are interested in finding out why one person gets arrested and another doesn't, even though they may both have committed the same crime, so you've lighted on a short passage titled 'Give a dog a bad name . . . '.

'Give a dog a bad name . . . '

Research in Britain and America has revealed several factors which influence police decisions. The first, and most obvious, is the nature of the offence. If it is serious, they are more likely to make an arrest regardless of the circumstances. But then, as we have seen, the vast majority of offences are not particularly serious. Second, if the offender has been in trouble before and is known to the police then, again, he is more likely to be arrested. This shows the effect of labelling. If a person has been labelled once, he is much more likely to be again, if he gets caught.

Finally, one of the most important factors is the offender's *demeanour* (the way he behaves towards the police when he is caught). If he is aggressive, cheeky, defiant or hostile, he is much more likely to be arrested than if he is polite and cooperative. However, there is also evidence that if he overdoes the politeness and helpfulness, this again is more likely to lead to arrest (perhaps because the police think he is 'trying it on'). The attitude of the *victim*, too, is important. If he wants the police to take action and is also polite and helpful, the offender will be more likely to be arrested.

With adolescents in particular, what the police often try to do is distinguish the really 'bad' characters from 'ordinary kids who happen to have got into trouble'. The actual offence is not necessarily a very good guide. Here, however, the police have come in for criticism. It is often suggested that they see particular groups — hippies, skinheads, coloured immigrants and so forth — as being specially troublesome and are biased against them. This criticism is, of course, difficult to prove. But it is certainly true that police attitudes towards particular social groups will influence their decisions about whether or not to make arrests.

Furthermore, if the police think that certain types of people are more likely to commit crimes, they will obviously look hardest amongst them when a crime has been committed. For example, if they think that a particular lower class slum area is full of criminals, they will police it heavily. Consequently, people from that area who commit crimes are more likely to get caught than offenders from other areas. And police records will show that the area *is* full of criminals! This does not, in itself, explain how 'criminal areas' arise. But it may have the effect of exaggerating how criminal they are compared with other areas.

The same thing can happen with social groups as well as areas. For instance, if the police think that members of adolescent gangs are likely to commit crimes, they are more likely to stop and question them. If this happens a lot, it is likely to make them hostile and defiant. And, as we have seen, this makes them more likely to be arrested. Moreover, being labelled as 'trouble-makers' or possible troublemakers, being stopped and questioned frequently when they have not done anything wrong, or being arrested when they have, can lead to a change in the way they see *themselves*. They may come to think of themselves as delinquents. In other words, they may accept the label and live up to it.

Even though it's been clearly written, there are some difficult ideas here. 'Active reading' is obviously called for. It looks as if most of the information you want is contained in the first two paragraphs and, as a way of understanding them, you could set about making a set of notes in the conventional way.

Police decisions based on:
1 the nature of the offence;
2 the offender's past record;
3 the offender's 'demeanour';
4 the attitude of the victim.

But the problem with a set of notes like this is that, whilst they are effective as a way of *recording* what you've already understood, they're not much use if you couldn't make sense of the passage in the first place. What you need is an approach that gets you closer to the text while you are actually reading it, one which helps you sort it out. If, instead of making notes, you had circled key words or phrases and underlined important points, then the passage might have looked like this when you had finished:

'Give a dog a bad name . . . '

Research in Britain and America has revealed several factors which influence police decisions. ①The first, and most obvious, is the nature of the offence. If it is serious, they are more likely to make an arrest regardless of the circumstances. But then, as we have seen, the vast majority of offences are not particularly serious. ②Second, if the offender has been in trouble before and is known to the police then, again, he is more likely to be arrested. This shows the effect of labelling. If a person has been labelled once, he is much more likely to be again, if he gets caught. ③Finally, one of the most important factors is the offender's *demeanour* (the way he behaves towards the police when he is caught). If he is aggressive, cheeky, defiant or hostile, he is much more likely to be arrested than if he is polite and cooperative. However, there is also evidence that if he overdoes the politeness and helpfulness, this again is more likely to lead to arrest (perhaps because the police think he is 'trying it on'). ④The attitude of the *victim*, too, is important. If he wants the police to take action and is also polite and helpful, the offender will be more likely to be arrested.

With adolescents in particular, what the police often try to do is distinguish the really 'bad' characters from 'ordinary kids who happen to have got into trouble'. The actual offence is not necessarily a very good guide.

Having done that, you may then be in a better position to make notes if you still need to. Alternatively, you could represent the ideas as a diagram:

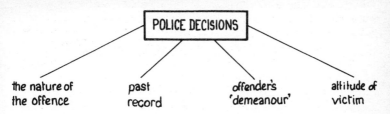

POLICE DECISIONS

the nature of the offence   past record   offender's 'demeanour'   attitude of victim

The advantage of this method is that because the *relationship* between different ideas is shown *visually*, it can sometimes be easier to understand. What's more, you can add further points at any stage without making so much of a mess that it becomes unreadable.

That example, which covers the first two paragraphs only, is fairly easy to understand, but the rest of the passage becomes a lot more complicated. It goes on to explain what effect there is on crime generally as a result of all these police decisions. A diagram to show this would have to be a lot more complex. Fortunately, the writer provides one himself, but by careful reading you could have arrived at this unaided:

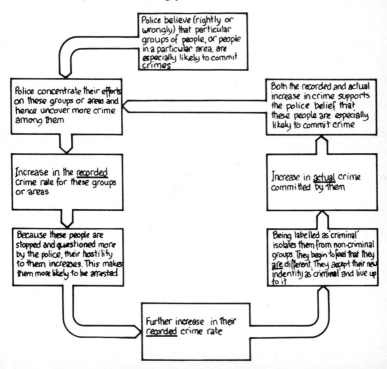

The diagram is in the form of a 'flow chart'. In other words, arrows have been used in order to show the direction for you to follow when you are reading it. It's particularly suitable as a way of representing this passage, which is all about *cause* and *effect*.

Don't fall into the trap of assuming that every piece of writing is like this, and that you can always make notes by using a flow diagram. Each time you settle down to make sense of a passage, you need to experiment with different approaches. Look again, for example, at two more paragraphs. They've been placed side by side here in order to show how the writer is drawing a *comparison* between what happens in certain areas, and with certain social groups. He even uses the same words to provide a shape for each paragraph:

Furthermore, if the police think that certain types of people are more likely to commit crimes, they will obviously look hardest amongst them when a crime has been committed. For example, if they think that a particular lower class slum area is full of criminals, they will police it heavily. Consequently, people from that area who commit crimes are more likely to get caught than offenders from other areas. And police records will show that the area *is* full of criminals! This does not, in itself, explain how 'criminal areas' arise. But it may have the effect of exaggerating how criminal they are compared with other areas. The same thing can happen with social groups as well as areas. For instance, if the police think that members of adolescent gangs are likely to commit crimes, they are more likely to stop and question them. If this happens a lot, it is likely to make them hostile and defiant. And, as we have seen, this makes them more likely to be arrested. Moreover, being labelled as 'troublemakers' or possible troublemakers, being stopped and questioned frequently when they have not done anything wrong, or being arrested when they have, can lead to a change in the way they see *themselves*. They may come to think of themselves as delinquents. In other words, they may accept the label and live up to it.

If it was this comparison that you wanted to make notes about, a flow diagram wouldn't be very helpful. You would be much better advised to draw two columns and make your notes in that way:

1 slum area	1 social group
2 policed more heavily	2 stopped and questioned more often
3 offenders more likely to be caught	3 offenders more likely to be hostile and, therefore, arrested
4 'more' crime	4 'more' criminals

If you were actually working on a passage like this, you obviously wouldn't be able to draw out two paragraphs and place them beside each other. But you could still underline the key phrases in each one, and number them 1, 2, 3, 4, or draw connecting lines between them in order to demonstrate to yourself what they have in common.

It's worth stressing that these are only examples of techniques that you could use. The important thing is that you should not be timid when you're working on a text. Experiment with different approaches and, if necessary, devise some of your own.

SUMMARY

1 Work out what you want from a book before you start reading it.
2 With reference books, use the index and table of contents to help find your way around, then SKIM and SCAN.
3 When studying a shorter passage in depth, be an ACTIVE reader:
   (a) where possible, mark the text;
   (b) have a piece of paper with you to make notes (e.g. lists, diagrams).

Examination tips

However confident you are about reading, you won't get any credit for what you can do unless you are also good at working under examination conditions. If you are fortunate enough to be assessed entirely by coursework, the next section is superfluous.

The advice that follows is designed to help you tackle short-answer questions, since this is where exam technique is likely to make the biggest difference.

1 Before you start, remember to finish

'I didn't have time to finish,' isn't going to gain you any extra marks. It is essential that you complete the whole paper. If you find you are getting 'stuck' on one question, go on to the next one. Don't get 'bogged down' on one question, it wastes time.

2 Make sure you understand the passage

Read the passage at a moderate pace (don't rush) and try to enjoy it. Don't attempt to speed things up by reading the questions first. This won't save you any time and it will only confuse you.

3 Read the questions carefully

(a) If you are asked to look at lines 35–42, don't base your answer on line 31 — you will get no marks.
(b) If the question asks you to 'quote' or 'give examples from the passage', you should quote. At other times you should, as far as possible, use your own words.
(c) If you are asked to 'give examples from the passage', give the exact number asked for. You will not get extra marks for extra answers. (When the question does not ask for an exact number of examples, use the number of marks allocated to that question as a rough guide to the number of examples required.)

4 Make sure you know what kind of answer you are required to give

Basically, there are three kinds of questions, and they require different kinds of answers:

(a) factual questions, where you are asked to 'quote', 'give examples' or show 'what evidence is available in the passage';

(b) comment questions, where you are asked to 'comment' or 'give your opinion' on something in the passage (These questions require you to keep to the point. Don't invent information or ideas and experience for the people in the passage.);

(c) 'essay' or 'imaginative' questions, where you are asked to tell a story or imagine a conversation involving characters in the passage. (These questions require you to keep to the mood of the passage.)

5 Always answer in complete sentences

Rephrase the question to form the introduction to your answer. If, for example, the question asks: 'Why was Michael scared when he heard the noise?', you might begin 'Michael was scared because he thought . . . '.

6 Keep to the evidence in the passage

Base all your answers on the evidence before you. Don't invent. If you find yourself tempted to use the word 'probably', alarm bells should start to ring. It may mean that you can't find any evidence to support your view.

7 Never answer from memory

Always turn back to the passage when answering a question. You should never rely on memory alone.

8 Don't ramble

If the question asks for several pieces of information or evidence, you may find it helpful to number each point in your answer.

9 Keep in mind the value of each question

Don't write three lines on a question which offers twelve marks, or a side on a question which only offers three.

10 Finally: Remember that you must finish.

SECTION 5
Oral communication

INTRODUCTION

Your first problem with this part of the syllabus will be to sort out what it is called. Don't get confused when teachers refer to it, almost at random, as 'spoken English', 'talk', 'the oral' or 'oral communication'. In the past, it's been all of these things. 'Oral Communication' is the term that is used in the National Criteria, and that is what it's likely to be called on your certificate.

You will also find that there are some important differences between this and other sections of the syllabus. The main one is that although 'Oral Communication' is compulsory, it will be 'separately assessed and will appear as an endorsement on the certificate, using a different, shorter grade scale'. What this means is that your grade for English (from A to G) will be based on your performance in the coursework or exams that cover reading and writing. But you won't actually be awarded a certificate unless you have also passed in 'Oral Communication', for which you will be graded between 1 and 5. For both parts of the exam, there will be an 'Ungraded' category. The final grade on the certificate, then, will be a letter *and* a number, C3, for example, or D2.

The reasons for this are not worth going into at great length, but it is just worth recalling the anxieties that many of the exam boards have about assessing something that can't be written down on paper, and adding that doubts have been expressed in the past about the sense of using only one grade to cover two skills in which people can show widely different levels of ability.

Perhaps the best way to introduce this part of the syllabus is to tell you what it's *not* about.

The first thing that it is not about is teaching people to speak 'properly' by ironing out regional accents or dialect. Most of the syllabuses will include reference to 'audibility, intelligibility and expressiveness' (NEA) and you need to be quite clear in your own mind about the point at which the way in which you talk can no longer be heard or understood by somebody else, but you will not be penalised for using a local accent.

Neither is it about 'public speaking'. You won't be expected to master the formal rules of debate or deliver formal speeches, even though plenty of opportunities should be provided for discussion.

Finally, it is not about acting. You won't have to *perform*, even though there may well be occasions on which others will be listening to what you have to say.

'Oral Communication' will be none of these things, because it is concerned with *what* you say as well as the way you say it, and because it covers a *variety* of situations in which speech is used.

This description is designed to allay the fears that you may have about 'Oral Communication'. Every teacher who has ever worked on a syllabus with an oral element knows that there are some pupils who become so nervous that they simply cannot bring themselves to talk in class. If you feel like that, you might be tempted to react by dismissing this part of the exam as pointless or stupid or not what English is about. One way of being more positive about it is to try to understand why oral communication is important and why it needs to be taught. Take 'discussion'. Don't imagine that it is just a pleasant way of passing the time. Often, it can be a vital part of the process of arriving at decisions or solving problems. Vital, because several people agreeing about something are more likely to be right than one person deciding on his or her own. If you find this hard to believe, try the following activity:

Lost at Sea

Instructions: You are adrift on a private yacht in the South Pacific. As a consequence of a fire of unknown origin, much of the yacht and its contents have been destroyed. The yacht is now slowly sinking. Your location is unclear because of the destruction of critical navigational equipment and because you and the crew were distracted trying to bring the fire under control. The best estimate is that you are approximately one thousand miles South South West of the nearest land.

Below there is a list of fifteen items that are intact and undamaged after the fire. In addition to these articles, you have a serviceable rubber life-raft with oars, large enough to carry yourself, the crew, and all the items listed below. The total contents of all survival pockets are a package of cigarettes, several books of matches and five one-dollar bills.

Your task is to rank the fifteen items below in terms of their importance to your survival. Place the number 1 by the most important item, the number 2 by the second most important, and so on through to number 15, the least important.

sextant
shaving mirror
twenty-litre can of water
mosquito netting
one case of US Army C rations
map of the Pacific Ocean
seat cushion (floating device approved by the Coast Guard)
small transistor-radio
shark repellent
two square metres of opaque plastic
one litre of 160 proof Puerto Rican Rum
five metres of nylon rope
two boxes of chocolate bars
fishing kit
ten-litre can of oil/gas mixture

You can only go on to the next stage if three or four of your friends have also completed their own individual lists of priorities. When that's done, get together and agree on the choice that you would make as a group. On page 81 is printed the order which would be recommended by experts in survival. With it are instructions about how to score what you have done.

Nine times out of ten the group mark is lower than the individual mark, indicating that, as part of a group, you got closer to the opinion of the experts than you did by working on your own. That's why discussion is important. By itself, however, it doesn't make a complete course in oral communiation. Just as there is variety in writing, so there is in talk.

You can get some idea of what this means by thinking about all the different kinds of situations in which you use spoken English. The National Criteria point out that some are more 'formal' than others, and that is a good starting-point for drawing up a list:

gossip/chat	job interviews
conversation with one other person	committee work
talk in small groups	speeches or talks
discussion	lectures
interviews	debates
reporting back	public enquiries
meetings	

It wouldn't be difficult to make this list a good deal longer. You can probably think of other examples yourself. The problem that the exam boards face, however, is the opposite one. They have to decide how to make it shorter, since no school syllabus could possibly hope to cover everything.

As with everything else in the GCSE, decisions like this about where 'Oral Communication' should begin and end must take account of the National Criteria, which say two things quite clearly about this part of the syllabus:

Emphasis should be placed on the inter-relatedness of listening and speaking . . .

and

For assessment purposes there must be some aim or goal, not simply a random exchange of views . . .

Taking them one at a time, you should be able to see how they help to make that list of possible activities more manageable:

Emphasis should be placed on the inter-relatedness of listening and speaking . . .

1 Listening comprehensions, in which you work from a sound tape, answering questions about what you've heard, are *out* because they test listening separately.
2 Talks and readings from a book, if used as the only test of oral communication, are *out* because they test speaking separately.
3 Interviews of various kinds, discussion and conversation are *in*, because they link listening and speaking naturally together, into one activity.

For assessment purposes there must be some aim or goal, not simply a random exchange of views . . .

1 Anything that might be described as 'gossip' or 'chat' is *out*.
2 Not all forms of discussion are equally acceptable. *In* is the sort of discussion where a number of people meet to solve a problem, *out* is the kind of casual chat in which opinions are exchanged but never challenged.

In one sense, none of this need worry you unduly. It's your teacher who will have to interpret the syllabus and decide what is allowed and what is not. If you understand what's going on, however, you are much more likely to be able to see what will gain you high marks and what won't. It will help you judge what to do when you go into your classroom one morning and find that you are going to be expected to do some talking.

sextant	15
shaving mirror	1
twenty-litre can of water	3
mosquito netting	14
one case of US Army C rations	4
map of the Pacific Ocean	13
seat cushion (floating device approved by the Coast Guard)	9
small transistor-radio	12
shark repellent	10
two square metres of opaque plastic	5
one litre of 160 proof Puerto Rican Rum	11
five metres of nylon rope	8
two boxes of chocolate bars	6
fishing kit	7
ten-litre can of oil/gas mixture	2

First compare your individual decision with the 'correct' answer by taking each item and subtracting the lower mark from the higher one. If, for example, you put the sextant (correct place: **15**) in seventh place, you would score 8(15 − 7 = 8). Add up the scores for each item and you will arrive at your individual mark. Then do the same with the group choice for each item, in order to produce a group mark. Now turn back to the point you had reached on page 79.

ACTIVITIES

Some activities are likely to be fairly commonly used by teachers and they are worth a closer look.

The talk

The old CSE exams often included a section in which you were expected to take a topic of your own choice and talk about it for five minutes or so, then answer questions from the class. In the GCSE there are likely to be some changes to this format so that any talk you deliver won't be expected to stand on its own, but will be treated as the introduction to a discussion or interview.

Here's a description, taken from the syllabus for one of the trial GCSE exams run by the Northern Examining Association, of what teachers should expect of an 'above average' talk. Whilst this shouldn't be treated as the last word on the subject, it provides a useful guide to some of the things that you need to think about:

Above average talks should provide evidence of careful preparation;

recognisable shape and structure; knowledge of subject; confidence and commitment; good eye contact; mature command of language and phrasing and delivery; firm audibility; clarity of pronunciation; intelligent use of visual materials, if used; creation of interest; ability to handle and answer questions. Those listening probably will have enjoyed the talk and felt at the end that their understanding of the subject has been enlarged.

What can you learn from this? The first thing is that if you are going to do any kind of uninterrupted talking, you can't expect to do it well without preparation. It's just not possible to keep in your head at the same time all the things that you need to be thinking about:

By making notes, you can eliminate some of these questions before you open your mouth, and that leaves you better able to concentrate on your audience in order to judge whether they are actually listening to you or not. This exam board warns, however, that: 'a well-written essay badly (or well) read does not qualify as a talk. It can receive a mark only in the "below average" category'. In fact, it's no good writing out everything you want to say beforehand because it's very difficult to guess, in advance, how different something will sound when it's spoken. You may, for example, suddenly realise that you are trying to explain a rather complicated point. In speech, it's quite acceptable to have another stab at it if your first attempt seems to be greeted by a row of blank expressions. If you're reading from a script, however, it's very difficult to break off and try again, without losing your place and finding that everything starts to fall apart.

The description of an 'above average' talk also suggests that it's important to be able to establish a relationship with your audience. It mentions 'good eye contact', for example, because this is one sign of whether you are in control of the situation. It can make the difference between actually communicating with a group of people and talking into thin air. The references to 'audibility' and 'creation of interest' are there for similar reasons. It's your job to make people feel that what you have to say is worth taking in. It's no good assuming that they'll be listening just because you are talking.

There are a hundred and one ways of achieving this, all of which are easily forgotten when you actually find yourself in the situation. You'll never learn to do it well, however, if you turn it into an ordeal that has to be completed as quickly as possible. Enjoy the opportunity that it provides to make you the centre of attention.

You are most likely to be able to do all this if you are feeling confident. Indeed, the exam board mentions 'confidence and commitment' as something which teachers need to be on the look out for. Confidence comes from knowing that, whatever else happens, you won't dry up completely. Two things can help here. First of all, if you have the opportunity, practise at home in front of somebody you can trust. Secondly, try to choose situations or topics on which you feel you are an 'expert'. Other people will only respond to what you are saying if you know what you are talking about. Finally, if you are *introducing* a discussion or conversation, you need to make sure that that is what you are actually doing. It's no good having the last word on the subject, and then being surprised when nobody else has anything to say. Use your introduction to raise matters on which you'd be interested to hear other people give their opinions. In other words, don't merely reel off a string of facts without commenting on them. Any information you provide should only be there in order to stimulate a good discussion afterwards.

Reading aloud

In the same category as the delivery of a talk is reading aloud. It's not acceptable as the *only* test of oral communication, but it can still form part of an overall assessment and may, like a talk, be used as the preparation for a discussion. The traditional test of reading aloud goes something like this:

Candidates read a prepared passage of about 500 words from a book, newspaper or magazine of their choice. They introduce the article or extract before reading and answer questions after the reading. (NEA)

It looks easy, doesn't it? There are none of the problems presented by the talk of working out what you're going to say, or worrying about whether the words will come out once you've started. It's all down there on the page and it simply has to be read out.

Don't be misled. In many ways it's far more difficult to do a good reading than a good talk. The problem is that you aren't going to get any credit for what somebody else has written, so all the marks that are available will be awarded for the *way* in which you read. Here, for example, is a description, again from the NEA, of a 'good reading':

The reading itself should have a firmness and clarity. The significance of character, incident and feeling should be suggested by attention to selective emphasis, changes in pace and tone and intelligent use of the voice. A degree of eye contact should be attempted.

Look at the key words: 'clarity', 'changes in pace and tone', 'use of the voice'. They are all concerned with the ability to control and alter your voice in order to reflect the meaning of the passage. You've got to be able to do a lot more than merely sound the words on the page. Indeed, fluent reading is taken for granted. Before venturing on a reading, then, you should ask yourself the following questions:

1 Am I able to glance ahead while I'm reading in order to see what is coming next and make allowances for it?
2 Can I look at the passage, read ahead, and remember it well enough to be able to look up at the people I am reading to?
3 Do I know enough about punctuation to be able to stop and start in the right places and demonstrate with my voice that that is what I have done?
4 Can I change the sound of my voice in order, for example, to indicate different characters speaking to each other?
5 Can I vary the speed at which I read from paragraph to paragraph?

Don't imagine that all these things will take care of themselves, they have to be learned. If you want to see how good you are, try making a tape-recording and listening to yourself. If you want a shock, compare it with a similar recording taken from a suitable radio programme.

Discussion

It's quite clear that discussion, in some form or other, is going to be the single most important element in 'Oral Communiation'. It's more difficult to be precise about exactly what form it will take in your school. Part of the reason for this is that the word 'discussion' covers a lot of different activities, ranging from a conversation between two people to a full-scale meeting. Consider, for example, the different demands made by these two situations:

A The class has been divided into pairs. Interviews are taking place about part-time or Saturday jobs. The aim is to discover as much as possible about a particular job in order to decide whether to apply for it if a vacancy occurs.

B The class has been asked to produce a magazine. An editorial committee of eight pupils has been formed with the job of agreeing on how to design the layout of the magazine and what articles should be included in it.

If you were involved in **A**, you'd need to be able to work well together, helping each other out by asking the right kinds of questions and by giving answers that really conveyed an impression of what the job was like. You probably wouldn't find it too difficult to know when to speak and when to listen because, between two people, it's fairly easy to work that out. But you wouldn't find out much if you simply ran down a list of questions that you had decided on beforehand. The real skill of this kind of interview lies in being able to listen carefully, and ask questions that arise out of what you've heard.

The second situation would be quite different. Again, you would need to be able to work well with other people, but this time it wouldn't be quite so easy. It is quite likely that you would all have very different views about what kind of magazine you wanted to produce. Unlike the interview about part-time work, this could be a discussion in which everybody involved was pulling in different directions. The key to success in that kind of situation lies in your ability to put your point of view at the right moment, and accept the majority verdict if you can't persuade anybody else to agree with you. But even deciding on what 'the right moment' is can be very difficult. It's no good simply blurting out that you think there should be a problem page when everybody else is discussing the design of the front cover, but that's always the temptation when a good idea occurs to you and you're worried you'll forget it. You also need to be able to help other people in the group. Sometimes that can be a matter of encouraging somebody who is feeling very hesitant, sometimes it means challenging somebody else who is saying too much, sometimes it means looking for a compromise between two different points of view. Finally, you've got to have something worth saying yourself and you have to be able to express it clearly.

The first task, then, when you are involved in discussion, is to work out what kind of activity it is, and what is expected of you. Just how much of a tightrope this can be is hinted at in the Guidance for Teachers given by the NEA:

The candidate who dominates without regard for the opinions of others is to be censured almost as much as the one who presents a prepared

statement, sits back and plays no further part. A candidate who makes no discernible contribution can score no marks.

Having decided what approach you need to take, whether it's to ask questions, to give information, to express an opinion or to ensure that a decision is reached, you've then got to find the best way of going about it. The words you use to express your ideas will always affect the way in which other people respond to them, sometimes with quite far-reaching consequences. It's possible, for example, to kill a discussion with a badly-chosen phrase or a thoughtless comment. Take disagreements. More than anything else, the direction that a discussion takes will be influenced by the language people use to contradict each other. Here are six phrases that are commonly used in speech to introduce a different point of view:

'Come off it!'
'But haven't you forgotten that . . . '
'You must be joking!'
'Don't be stupid!'
'I don't agree!'
'The way I see it is like this . . .'

There are some situations in which all of these expressions would be perfectly acceptable. Imagine, though, that you were involved in a reasonably formal discussion in class in a group with other pupils whom you didn't know particularly well. Which of these phrases would *encourage* others to contribute to the conversation, which would *discourage*? It's this kind of decision that you need to be making all the time if you are to become skilful in discussion.

Role-play

In order to provide a greater variety of situations for oral communication, some teachers use a technique called 'role-play'. This involves putting yourself in an imaginary situation and finding out how you would cope with it. Some examples might help:

1 You are applying for a job working as a clerical assistant in an office. Conduct the interview.
2 You've been out to a party, your parents say you must be back by midnight. It's one o'clock as you open the front door and they have waited up for you. What happens?
3 An old man lives in the house next door, which has been condemned by the local council. He's refusing to move. Can you persuade him?

But, you might protest, 'Oral Communication' isn't meant to be about acting. That's a reasonable point, and it's one of the reasons why a lot of teachers may not use this technique. However, although there are some similarities between role-play and acting, there are also some important differences. It's possible to illustrate this by looking a bit more closely at one of the situations described above. If you were taking on the role of the old man in the condemned house (not part of this particular question), it wouldn't be necessary for you to behave like an old man. You wouldn't be expected to pretend that you had arthritis or to speak in a quavering voice. Nor would you be expected to create a character unlike your own. What role-play is designed to do is to look at what *you* would say or do in those circumstances. A teacher or examiner would be interested in whether you could find the right words, not whether you could put on an Oscar-winning performance. If you could express, say, the fears that an old man might have about where he would go if he were evicted and the problems this would present in picking up the threads of his life, then you would be likely to score high marks. Similarly with the other role, the person persuading him to move. There are a number of approaches you might take, but you would obviously have to find a way of gaining the old man's confidence whilst still pointing out that the house is a health hazard. Whatever you said, you would have to choose your words carefully, and that's what you would be marked for.

It all comes back to the point that was made earlier about how the way in which you express something affects what other people say and do. Even more than discussion, role-play asks you to think not just about what you want to say, but also the words you are going to use to say it.

METHODS OF ASSESSMENT

Although the National Criteria don't rule out the possibility of 'Oral Communication' being tested by a visiting examiner, most of the exam boards have decided against doing it that way. The person who will probably be responsible for your grade will be your teacher, and the job of the exam board will be to check that teachers are awarding marks in the right way. What this means is that your grade will depend upon what you do during the course and, in theory at least, everything you do could make a contribution. It's likely, though, that teachers will arrange particular occasions when they can concentrate on assessment, and they will let you know when this is happening. The difficulty that they will face during the early stages of the GCSE is that the grading system has never been used before and so nobody has any experience of how it will work. Descriptions of what the grades will mean have been produced for teachers and they might be of some help to you. Although the grades go from 1 to 5, the two examples that are reprinted here cover only grades 2 and 4.

GRADE 2

The candidate can be expected to have demonstrated competence in:

6.6.1 understanding and conveying both straightforward and more complex information;

6.6.2 ordering and presenting facts, ideas and opinions with a degree of clarity and accuracy;

6.6.3 evaluating spoken and written material and highlighting what is relevant for specific purposes;

6.6.4 describing and reflecting upon experience and expressing effectively what is felt and what is imagined;

6.6.5 recognising statements of opinion and attitude and discerning underlying assumptions and points of view;

6.6.6 showing sensitivity in using a range of speech styles appropriate to situation and audience;

6.6.7 speaking clearly and coherently with appropriate tone, intonation and pace.

GRADE 4

The candidate can be expected to have demonstrated competence in:

6.5.1 understanding and conveying straightforward information;

6.5.2 presenting facts, ideas and opinions in an orderly sequence;

6.5.3 selecting and commenting on spoken and written material with some sense of relevance;

6.5.4 describing experience in simple terms and expressing intelligibly what is felt and what is imagined;

6.5.5 recognising statements of opinion and attitude;

6.5.6 using some variation in speech style according to situation and audience;

6.5.7 speaking audibly and intelligibly with some sense of appropriate tone, intonation and pace.

It's very difficult to put this kind of thing down on paper and you certainly shouldn't take it too literally. However, it does give you some idea of what your teachers will be taking into account when they are marking you.

For a more complete understanding of what each item in the grade description means, see if you can match them up with each of the following tasks (bearing in mind that none will fit precisely). You might feel, for instance, that the last one, making recommendations for a good night out, most closely fits item **6.6.2**: 'ordering and presenting facts, ideas and opinions'.

Could you:

1 give a stranger a clear set of directions;
2 find a way of dealing with a customer if you were serving in a shop;
3 find what to say when you were taken to visit an elderly relative whom you hadn't seen for years;
4 describe a rock concert to a friend who hadn't been able to go;
5 read a pamphlet on supplementary benefit and explain to somebody what their entitlement was;
6 explain how to work a sewing machine;
7 recommend places to go for somebody who didn't know your area and wanted a night out?

These are merely examples, they wouldn't make for very successful classroom activities, but they put some flesh and blood on the bare bones of the grade descriptions.

There's one final point that needs to be made about oral communication. You probably associate school mainly with writing. A lesson in which talk is not just allowed but positively encouraged might not seem like work. You should by now appreciate that there is just as much for you to learn about oral communication as there is about reading or writing, and that, in many ways, you need more self-discipline when working in a group than you do when working by yourself. If you've grasped that basic fact, then you're half way to success already.

SECTION 6

Literature

INTRODUCTION

In addition to 'English', you may be taking a separate exam in 'English Literature'. The introduction of the GCSE means that the syllabus you will be following is significantly different from almost all of the syllabuses that have been available in the past; and, as with 'English', change was long overdue.

In describing what's new about GCSE, it's convenient to start with the National Criteria:

The assessment objectives in a syllabus with the title ENGLISH LITER-ATURE must provide opportunities for pupils to demonstrate, both in the detailed study of some literary texts *and in wider reading*, their ability to:

1 **acquire first hand knowledge of the content of literary texts;**
2 **understand literary texts, in ways which may range from a grasp of their surface meaning to a deeper awareness of their themes and attitudes;**
3 **recognise and appreciate ways in which writers use language;**
4 **recognise and appreciate other ways in which writers achieve their effects (e.g. structure, characterisation);**
5 **communicate a sensitive and informed *personal response* to what is read.**

Most of that would not be out of place as an introduction to one of the old 'O' level or CSE syllabuses which led to traditional exams in which questions were asked about three or four books and answered from memory.

If they were proposed for GCSE, however, those 'O' level and CSE syllabuses would be rejected on two counts. Firstly, they did nothing to encourage 'wider reading', and secondly, they were woefully inadequate when it came to assessing 'a sensitive and informed personal response'. Put together, these two requirements make almost inevitable the third major change that will affect you. Instead of sitting an exam, you will be expected to produce a folder of coursework for part, if not all, of your course. The reason for this is that it's difficult to find any other way of assessing 'wider reading', which by definition covers more than a few set books, or 'personal response', which is hard to conjure up under exam conditions.

It's not easy to explain what all this will mean in practice, because different teachers will interpret the new syllabuses in different ways. Imagine, however, that you were studying *Kes*, by Barry Hines. It's a novel that has often been set in the past both for 'O' level and CSE, and it will probably be used again. The teacher plans to work methodically through the book so, on the first day, you start at chapter one:

There were no curtains up. The window was a hard edged block the colour of the night sky. Inside the bedroom the darkness was of a gritty texture. The wardrobe and bed were blurred shapes in the darkness. Silence.

Billy moved over, towards the outside of the bed. Jud moved with him, leaving one half of the bed empty. He snorted and rubbed his nose. Billy whimpered. They settled. Wind whipped the window and swept along the wall outside.

Billy turned over. Jud followed him and cough-coughed into his neck. Billy pulled the blankets up round his ears and wiped his neck with them. Most of the bed was now empty, and the unoccupied space quickly cooled. Silence. Then the alarm rang. The noise brought Billy upright, feeling for it in the darkness, eyes shut tight. Jud groaned and hutched back across the cold sheet. He reached down the side of the bed and knocked the clock over, grabbed for it, and knocked it farther away.

'Come here, you bloody thing.'

He stretched down and grabbed it with both hands. The glass lay curved in one palm, while the fingers of his other hand fumbled amongst the knobs and levers at the back. He found the lever and the noise stopped. Then he coiled back into bed and left the clock lying on its back.

'The bloody thing.'

He stayed in his own half of the bed, groaning and turning over every few minutes, Billy lay with his back to him, listening. Then he turned his cheek slightly from the pillow.

'Jud?'

'What?'

'Tha'd better get up.'

No answer.

'Alarm's gone off tha knows.'

'Think I don't know?'

He pulled the blankets tighter and drilled his head into the pillow. They both lay still.

'Jud?'

'What?'

'Tha'll be late.'

'O, shut it.'

'Clock's not fast tha knows.'

'I said S H U T I T .'

He swung his fist under the blankets and thumped Billy in the kidneys.

'Gi'o'er! That hurts!'

'Well shut it then.'

'I'll tell my mam on thi.'

Jud swung again. Billy scuffled away into the cold at the edge of the bed, sobbing. Jud got out, sat on the edge of the bed for a moment, then stood up and felt his way across the room to the light switch. Billy worked his way back to the centre and disappeared under the blankets.

'Set t'alarm for me, Jud. For seven.'

'Set it thi sen.'

'Go on, thar up.'

Jud parted Billy's sweater and shirt, and used the sweater for a vest. Billy snuggled down in Jud's place, making the springs creak. Jud looked at the humped blankets, then walked across and pulled them back, stripping the bed completely.

'Hands off cocks; on socks.'

For an instant Billy lay curled up, his hands wafered between his thighs. Then he sat up and crawled to the bottom of the bed to retrieve the blankets.

'You rotten sod, just because tha's to get up.'

'Another few weeks lad, an' tha'll be getting up wi' me.'

He walked out on to the landing. Billy propped himself up on one elbow.

'Switch t'light out, then!'

Jud went downstairs. Billy sat on the edge of the bed and re-set the alarm, then ran across the lino and switched the light off. When he got back into bed most of the warmth had gone. He shivered and scuffled around the sheet, seeking a warm place.

In six months' time you'll be sitting in an exam room facing a question that might look something like this:

(a) **By describing in detail Billy's experiences at school, explain why Billy is not a successful pupil.**

(b) **Why is the bird Kes so important in the story? Refer closely to what happens to support your answer.**

You won't have your copy of the book with you, so everything you put down in your answer will have to have been memorised. Both you and your teacher know this, so what happens during the lesson? As far as the teacher is concerned, the most important thing is that you end up with a set of notes you can learn in the weeks immediately before the exam so that you are fully prepared for whatever questions are thrown at you.

Most of the lesson, then, will be spent writing down what you are told about the book. You might start with a plot summary, on the grounds that the first thing you need to be sure about is who the characters are, and what they are doing. (Billy and Jud are brothers. Billy is still at school, but will soon be leaving. Jud is already working, at the local pit, hence the early morning start. The book begins as they get up one morning and the story covers the events of one day in their lives.) Your teacher might then go a bit deeper and start to talk about what impression the opening pages give of the life they lead. The family is obviously poor, there are no curtains at the window, Billy and Jud have to share a bed despite their age, and Billy at least seems to be perpetually cold. The way they talk, in Yorkshire dialect, and the fact that Jud is working an early morning shift also help to establish that this is a Northern working-class family, and so we begin with a lot of expectations about how they live. Mention is made of the way in which the wind 'whipped' the window, and you get the feeling that, like the weather, life for Billy is pretty bleak.

Moving on to the two characters, you would probably have your attention drawn to the fact that Billy is the victim and Jud the bully. Jud is violent, both in what he says and what he does. Even in this short extract he thumps Billy in the kidneys twice and, for no reason other than to be spiteful, pulls the blankets off him. He has no respect for Billy or his possessions, and he is completely unhelpful, refusing to re-set the alarm or switch off the light. As far as Jud is concerned, Billy is just a nuisance. Billy, knowing when he's beaten, has devised subtle ways of trying to get what he wants. He reminds Jud about the clock because he wants to get him out of the bed and out of the house, but he doesn't dare say that. However, he isn't completely intimidated by Jud and he still answers back, even though he probably knows what the consequences will be. He seems to be a lad whose only concern is with surviving in a world where nobody's going to go out of their way to help him.

At the end of the lesson, if you had been listening attentively, you and your teacher might both feel that you now understood a bit more about the beginning of *Kes*. You've covered the *plot*, you've looked at the *characters* and, in the final comments about how Billy is a 'survivor', you've even started to touch on the *themes*. Over the weeks, your notes would build up into an impressive body of work. Indeed, in a survey of literature teaching in schools, the Inspectors reported that, 'a fifth year group in English had written 23000 words of dictated plot of *Far from the Madding Crowd*, and would have written many more when they had finished *Great Expectations*, already under way'. Put together, that's more than the whole of this book you are reading now.

Now think about the problems that these classes would face if, instead of sitting an exam at the end of it all, they were submitting folders of coursework which had to include work that communicated 'a personal response'. One of the exam boards (MEG) actually says that 'no credit may be given to copied or dictated work' and another (NEA) insists that every folder includes 'at least one piece of work which demonstrates the candidate's own unaided and spontaneous response'. No longer would it be sufficient to reproduce from memory everything that you had been told. The examiners wouldn't be very interested in whether you could relate the plot — after all, that's not very difficult to do if you've got the book in front of you. Neither would they be very interested in what the teacher thinks about it, since there's no great skill in copying out somebody else's views. Once those two approaches have been ruled out, all those lessons spent taking notes begin to look a bit unnecessary. A different way of teaching literature needs to be devised, one which will allow you to respond in your own way, so that what you write will be what *you* think.

Put in its most simple terms, teachers will have to stop providing the answers and start asking the questions. You'll have to stop copying things down and start working them out. Let's return to that lesson on *Kes*. Here's a passage a bit further on. Soon after Jud leaves the house, Billy gets up to do his paper-round before going to school. He's just arrived at the newsagent's, having walked because Jud has taken his bike:

> A bell tinkled as he entered the shop. Mr Porter glanced up, then continued to arrange newspapers into overlapping rows on the counter.
>
> 'I thought you weren't coming.'
>
> 'Why, I'm not late, am I?'
>
> Porter pulled a watch out of his waistcoat pocket and held it in his palm like a stopwatch. He considered it, then tucked it away. Billy picked up the canvas bag from the front of the counter and ducked under the strap as he slipped it over his head and shoulder. The bag sagged at his hip. He straightened a twist in the

strap, then lifted the flap and looked inside at the wad of news-
papers and magazines.

'I nearly wa' though.'

'What do you mean?'

'Late. Our Jud went to t'pit on my bike.'

Porter stopped sorting and looked across the counter.

'What you going to do, then?'

'Walk it.'

'Walk it! And how long do you think that's going to take you?'

'It'll not take me long.'

'Some folks like to read their papers t'day they come out, you
know.'

'It's not my fault. I didn't ask him to take it, did I?'

'No, and I didn't ask for any cheek from you! Do you hear?'

Billy heard.

''Cos there's a waiting list a mile long for your job, you know.
Grand lads an' all, some of 'em. Lads from up Firs Hill and round
there.'

Billy shuffled his feet and glanced down into the bag, as though
one of the grand lads might be waiting in there.

'It'll not take me that much longer. I've done it before.'

Porter shook his head and squared off a pile of magazines by
tapping their four edges on the counter. Billy sidled across to the
convector heater and stood before it, feet apart, hands behind his
back. Porter looked up at him and Billy let his hands fall to his
sides.

'I don't know, it's typical.'

'What's up, I haven't let you down yet, have I?'

The bell tinkled. Porter straightened up, smiling.

'Morning, Sir. Not very promising again.'

'Twenty Players.'

'Right, Sir.'

He turned round and ran one finger along a shelf stacked with
cigarettes. His finger reached the Players and climbed the packets.
Billy reached out and lifted two bars of chocolate from a display
table at the side of the counter. He dropped them into his bag as
Porter turned round. Porter traded the cigarettes and sprang the
till open.

'Than-kyou,' his last syllable rising, in time to the ring of the
bell.

'Good morning, Sir.'

He watched the man out of the shop, then turned back to
Billy.

'You know what they said when I took you on, don't you?'

He waited, as though expecting Billy to supply the answer.

'They said, you'll have to keep your eyes open now, you know, 'cos they're all alike off that estate. They'll take your breath if you're not careful.'

'I've never taken owt o' yours, have I?'

'I've never given you chance, that's why.'

'You don't have to. I've stopped getting into trouble now.'

Porter opened his mouth, blinked, then pulled his watch out and studied the time.

'Are you going to stand there all day, then?'

He shook the watch and placed it to one ear.

'Next thing I know, everybody'll be ringing me up, and asking why I can't deliver on time.'

Billy left the shop.

Before you start making notes, you need to be able to think about what should be in them. You might, for example, have to find answers to questions like these:

1 What sort of man is Mr Porter?
2 Do you like him?
3 How does he treat Billy?
4 How does Billy keep on the good side of him?
5 Does your attitude towards Billy change because he steals the chocolate?

The teacher's job becomes one of helping you to answer these questions effectively so that you are actually meeting the requirements of the exam. Remember, you've got to be able to go beyond an understanding of the 'surface meaning' to a 'deeper awareness'.

Ideally, you will eventually learn to ask *yourself* the questions, so that you can study a book. Try putting yourself in your teacher's shoes. A few pages further on there's a paragraph that describes one of the houses on Billy's paper-round. What questions would you want to ask the class about it?

At the side of the house, a grey Bentley was parked before an open garage. Billy never took his eyes off it as he walked up the drive, and when he reached the top, he veered across and looked in at the dashboard. The front door of the house opened, making him step back quickly from the car and turn round. A man in a dark suit came out, followed by two little girls in school uniform. They all climbed into the front of the car, and the little girls waved to a woman in a dressing gown standing at the door. Billy handed her the newspaper and looked past her into the house. The hall and stairs were carpeted. A radiator with a glass shelf ran along one wall, and on the shelf stood a vase of fresh daffodils. The car freewheeled down the drive and turned into the lane. The woman waved with the newspaper and closed the door. Billy walked back, pushed the letter box up and peeped through. There was the sound of running bath water. A radio was playing. The woman was walking up the stairs, carrying a transistor. Billy lowered the flap and walked away. On the drive the tyres of the car had imprinted two patterned bands, reminiscent of markings on a snake's back.

This account of how GCSE will alter the way in which 'English Literature' is studied is probably a bit misleading. For most candidates, it won't be a question of one approach or the other, but a combination of the two. It makes the point, though, that there are parts of the exam, particularly the 'wider reading' section, which will expect you to acquire the skills that are needed to read all books, rather than the information that can be gathered from just one or two.

THE EXAMINATION

For 'English Literature' the exam boards have provided a dazzling variety of different kinds of courses. The only things that you can be certain of are:
1 you will have to cover what are known as the three main *genres* of English literature: poetry, drama (plays) and prose (novels);
2 at least 20% of your marks will be awarded on the basis of course-work;
3 you will have to study some individual texts in detail;
4 you will have to do some 'wider reading'.
For full details of the scheme that you are following, you will have to rely on your teacher, who will be constructing your syllabus by picking and choosing from the alternatives overleaf:

CHOICE OF LITERATURE TYPES OF ASSESSMENT	'unseen' passage	set text	open choice of text
traditional exam			
'open book' exam			
coursework			

The terms used may need explanation. On the left-hand side is a list of methods of assessment. An 'open book' exam is one in which you are allowed to have with you in the exam the books that you are studying, so there's no need to commit large portions of them to memory. Along the top are described the different ways in which the literature you are studying might be selected. Texts can be chosen either by the Exam Board (set texts), by the teacher, or, as in the case of 'wider reading', by yourself (open choice). You may also have to sit an 'unseen' paper in which you will be expected to answer questions on a piece of writing that you come across for the first time in the exam room.

As you can probably imagine from all of this, there are a number of different kinds of questions that examiners ask about literature, and you need to prepare for them in different ways. The next section takes a look at some of the more common ones.

STYLE OF QUESTIONS

Two kinds of questions have traditionally been set in 'O' level and CSE exams — the general question about some aspect of the characters or themes in the set book, and the 'unseen' extract. In the last few years, however, one or two of the exam boards have become unhappy with this approach. The general question, they have argued, in which you have 40 minutes to comment on, say, the development of a major character in a Shakespearean play, is an invitation to be superficial. The 'unseen' extract, on the other hand, which allows you to respond in detail, is very demanding because there's never enough time available for you to become really familiar with the passage you are writing about. The solution that they have found is to devise exam papers in which you are expected to work on extracts *taken from one of the books that you have studied*. It's in this direction that GCSE is most likely to go, although all three kinds of questions will regularly be set, whether you are sitting exam papers, submitting a coursework folder or doing a combination of the two.

The general question

Here are some questions taken from the first sample papers produced for the GCSE:

1 What have you found to be enjoyable or interesting in a book or short story which you have studied? You may wish to consider characters, events, setting or ideas. **(LEAG)**

2 Write about a book or books you have read in which conflict occurs between people or between the individual and authority. Show how conflict develops. **(LEAG)**

3 (*Romeo and Juliet*)   The Nurse has agreed to be interviewed about the tragedy of Romeo and Juliet. What do you think she would say about her part in the sensational events that have rocked the city? You may write as the nurse if you wish. **(MEG)**

4 (*Journey's End*)   How does each of the officers try to come to terms with life in the trenches? Which of them do you think is most successful? Give reasons for you choice. **(MEG)**

5 (*To Kill a Mockingbird*)   Why do you think Sheriff Tate is determined not to reveal the full truth of the attack on the Finch children on their way home from the pageant? **(SEG)**

6 (*To Kill a Mockingbird*)   There are three fathers in this story – Atticus, Mr Radley and Bob Ewell – and each has his own way of bringing up children. Write a brief description of each man's attitude towards his children, commenting on how Atticus's attitude differs from those of the other two. What do you feel about the way these three men bring up their children? **(MEG)**

7 This selection of Science Fiction stories has no introduction. From your knowledge of the stories selected, write an introduction telling the reader what to expect of Science Fiction. **(SEG)**

Don't worry if you haven't read any of the books that are mentioned here. In understanding how the questions work, there's some advantage in not having your mind cluttered with ideas about how you would answer them. What you need to look at is not what they ask about *To Kill a Mockingbird* or *Romeo and Juliet* or *Journey's End*, but what makes them the kind of questions that they are.

Reading down the list you might think, 'But they're all completely different from each other. How can I possibly know what to expect? Don't panic. It's true that the questions set in GCSE 'English Literature' are likely to be much less predictable than in the old exams. But if you look again at those examples and try to work out whether they have anything in common with each other, you'll start to get some idea about what the examiners are looking for. Here's another, much simpler list of questions. Between them, they cover everything that was asked in the seven examples that we started with.

In the book you are studying:

Do you know what happened and why?
Were you affected by it?
Do you have an opinion about it?
Can you see how the writer did it?

Most of the questions that you have to tackle fall into one or other of these categories.

It's not hard to spot which of the questions from the sample exam papers are designed to find out whether you know what happened and why. Question 5 begins 'Why do you think Sheriff Tate . . .'. It's not uncommon for questions of this kind to start by asking 'Why?' That's because you're being asked to show that you can understand what makes the characters tick, to show that there are reasons for what they do. It's not always easy to work out the answer, of course. Sometimes, just as in life, you'll find that characters don't actually explain their motives, and that the writer isn't being particularly forthcoming either. It's when you're faced with that problem that you have to use your imagination. Look again at question 3. Although it doesn't actually start 'Why?', it's a similar kind of question. By asking you to put yourself in the position of the Nurse, it's trying to help you think about why she acts in the way she does. This is what people mean when they talk about *identifying* with characters. If you look at things from their point of view, you'll understand them better.

The only example of the kind of question that asks 'Were you affected by it?' is question 1. You might find it helpful to think about this as a *What?* kind of question, since that is the way in which they usually start – 'What have you found amusing/frightening/sad and so on?' This kind of question is a bit deceptive. It looks easy because all it seems to be asking you to do is to give a straightforward description of what happens in the book. Be warned, though. A description of the plot is not likely to earn you very high marks. The examiners also want to know *why* you found it enjoyable or interesting or amusing or frightening or sad. You've got to be able to think about your response and pin down what has sparked it off. Is it a particular piece of description? Is it a character? Is it something about the events that occur?

Questions 4 and 6 are different again. Instead of asking what you *think*, they ask what you *feel*. In other words, they want to know 'Do you have an opinion about it?' Questions like this come in all sorts of disguises, and you might find them particularly hard to recognise. The first one, question 4, is a *Which?* question. You have to decide which of the soldiers is most successful in coming to terms with life in the trenches. The phrasing in question 6 is a bit different, but even though it doesn't ask you to make a choice, it clearly asks you to pass judgement on the characters in a similar way. You can usually spot this kind of question by keeping an eye open for whether the wording is directed personally at you. Try and keep in your mind a clear distinction between whether a question is asking you to explain what happens in the book or decide what you think about it.

Finally, there are questions that ask 'Can you see how the writer did it?' In some ways, these are the most difficult questions to deal with. To answer all the others, you need to be involved in what you are reading. To answer these questions, *How?* questions, you need to be a bit more objective and analytical. You have to stand back from the book. Put another way, it's like the difference between driving a car and knowing how the engine works. The two examples of this kind are questions 2 and 7. Question 2 – 'Show how conflict develops' – has all kinds of clues in it about what is expected, most obviously in the phrase 'Show how'. Question 7 might give you pause for thought. It asks you to tell the reader 'what to expect of Science Fiction'. That means you need to have thought not just about a single Science Fiction story, but Science Fiction stories in general. You need to have a view about the genre.

It would be impossible to go through every question that might appear on an exam paper. The point is that you need to get into the habit of looking at all questions with a very professional eye. You need to learn the language of exam questions so that you know at once what is being expected of you, what is allowed and what is not.

The old GCE and CSE exams offered you far less choice. They tended to insist that everybody wrote about books in the same kind of way. With the new ways of asking questions, there will come new ways of marking them, and the examiners will be much more concerned to look at what you know, rather than what you don't know. They'll be looking for positive achievements. You are being given a chance, then, to show off, to use the questions to get across to the examiner your thoughts, feelings, opinions and judgements about the books you have been reading.

None the less, there'll still be expectations. There'll still be certain kinds of things that the examiners will want to see before handing out the top grades:

There . . . remain vast numbers of candidates who seem to come ill-prepared to cope with any but the simplest request for an account of the contents of the text. It has been stated frequently in examiners' reports during recent years that mere account is not sufficient to obtain a satisfactory grade unless it is accompanied by the comment or opinion that is usually invited as part of the question. To offer account where the question does not ask for it is a recipe for disaster.   (JMB)

This is taken from a report on an 'O' level examination, but you can be fairly confident that the examiners for the GCSE will become just as irritable if they feel you are doing no more than retelling the story.

You also have to avoid going to the opposite extreme. An essay in which there is no reference to the book you are studying is equally unlikely to win the examiner's approval. Even where you have been asked to express an opinion, you can't just say what you think. You've got to provide an argument that is good enough to convince somebody else, and that means going back to the text to find evidence.

In preparing yourself for questions like this, you've got to have a thorough knowledge of the book. To some extent you've also got to have 'pre-packaged' what you think about it. It's no good going into the exam room with a whole set of responses (as if you had read the book for pleasure), but no firm opinions. You'll find that under exam conditions it's very difficult to come up with new ideas. All you'll be able to do is to organise what you already know to suit the question you are answering.

The key to this lies in building up a set of notes on all the aspects of the book that seem to be important. It's a bad idea to rely wholly on notes that you've been given by somebody else, even if that person is your teacher, since you won't necessarily understand them all and you'll find it much more difficult to learn them. Start by deciding on a series of major headings – it might be characters and/or themes. Then gather together under each heading all the important points that you might want to make. It's at this stage that you can and should incorporate other people's ideas.

In sorting out the material you have gathered under each heading, you need to distinguish between important and less important ideas, and between ideas and examples, labelling each one. There are a number of recognised ways of doing this. Some people choose to use **A**, **B**, **C** for major headings. each of which might have sub-headings **1**, **2**, **3**. The sub-headings can be further divided up into **a**, **b**, **c**, and examples added: **i**, **ii**, **iii**. Another way to go about it is to use numbers only: **1**; **1.1/1.2/ 1.3**; **1.1.1/1.1.2/1.1.3** and so on. Whichever method you adopt, it's worth compiling your notes in this way as it provides a logical system for ordering your ideas, and forces you to be absolutely clear about what you think and what evidence you have to support your views.

The 'unseen'

At the other end of the spectrum from the general question is the 'unseen' extract. The 'unseen' asks you to look in close detail at a very short piece of writing. Since the chances are that it will be entirely new to you, you can't bring any prepared answers to it. What it tests is whether you have acquired the *skills* that reading literaure requires. Here's an example:

Instructions to candidates:

You are advised to spend about 40 minutes on this paper.

Ninetieth Birthday

You go up the long track
That will take a car, but is best walked
On slow foot, noting the lichen
That writes history on the page
Of the grey rock. Trees are about you
At first, but yield to the green bracken,
The nightjar's house: you can hear it spin
On warm evenings; it is still now
In the noonday heat, only the lesser
Voices sound, blue-fly and gnat
And the stream's whisper. As the road climbs,
You will pause for breath and the far sea's
Signal will flash, till you turn again
To the steep track, buttressed with cloud.
And there at the top that old woman,
Born almost a century back
In that stone farm, awaits your coming;
Waits for the news of the lost village
She thinks she knows, a place that exists
In her memory only.

                    You bring her greeting
And praise for having lasted so long
With time's knife shaving the bone.
Yet no bridge joins her own
World with yours, all you can do
Is lean kindly across the abyss
To hear words that were once wise.

R.S. Thomas

(a) We are told that it's a long climb to the top of the track, and so steep that you will pause for breath. Yet, though it 'will take a car', the track 'is best walked on slow foot'. If you were to drive up, what sights and sounds would you miss? What other reasons for walking 'on slow foot' does the poem make you feel you might have?

Why, do you think, does the writer use more than half the poem (14 out of 26 lines) to describe the journey to the 'stone farm', before he even mentions the 'old woman' who lives there? Link your answer to the actual words of the poem whenever you can.

(b) Though there is no direct description of the 'old woman' whose ninetieth birthday Thomas is celebrating, it is possible for us to gather, from his poem, certain impressions. What are your impressions of her? Remember to say which words and phrases suggest them to you.

(c) What do you see as the gulf ('abyss') that Thomas feels there is between the old woman and himself? How does he make you feel the width of this gulf?

(d) Write about anything else in the poem that has particularly interested you. (This may include words, phrases, striking images, a feeling, an attitude — or anything else that has struck you.)

(e) How does the poem leave you feeling — sad? glad? — or what?

The questions on this paper aren't completely typical, they're much better written than in some examples that could have been chosen, but they are sufficiently representative to give you the general idea.

It's worth noting, first of all, the way in which the questions are organised. The first three are about particular parts of the poem and they are arranged in a logical order, so that **(a)** is about the beginning of the poem and **(c)** about the end, whilst the last two questions cover the poem as a whole. The questions also become more difficult as you work through them. The first question in **(a)** simply asks for information, what the National Criteria call an understanding of 'surface meaning'. The second part of the question goes a bit further ('a deeper awareness'). It asks you to look for reasons that aren't actually stated but which are implied. The third question in **(a)** is aimed at seeing whether you can understand the techniques used by the writer. Again, this is covered in the National Criteria, which refer to 'the ways in which writers use their effects'. Question **(b)** also asks you to 'read between the lines' — it even uses the word 'impressions', though it lets you know that those impressions should be based on 'words and phrases' in the poem which 'suggest them to you'. Question (c) expects you to 'appreciate ways in which writers use language' (National Criteria) by focusing your attention closely on a few particular words. The final questions are designed to encourage a 'personal response'.

With these hints in mind, think about what kinds of answers you might give.

If you were to drive up [the track], what sights and sounds would you miss?

First of all, you would need to refer to the lichen on the rock, the trees and the bracken mentioned in lines 3–6. Later on, however, R.S. Thomas talks about the view from the steep track of clouds nearby and the sea in the distance. This shouldn't be ignored. The question also asks about 'sounds' – the blue-fly, the gnat and the stream are all mentioned, and R.S. Thomas says that on warm evenings you could hear the 'nightjar'. There's not much room for disagreement about what's required here. The second part, however, is a bit more difficult.

What other reasons for walking 'on slow foot' does the poem make you feel you might have?

What this is probably getting at is the atmosphere that is created in the first verse. Three things stand out – it is very hot, very still and very quiet. It's almost as if time were standing still. If you couple that with the phrase about 'the lichen that writes history on the page of the grey rock', you can get some sense of how the walk is important as a way of allowing you to go slowly back into the past to meet the old woman at the top of the hill. A car would take you there too quickly, bringing the present with it.

Why, do you think, does the writer use more than half the poem to describe the journey to the 'stone farm' before he even mentions the 'old woman' who lives there?

The third question is aimed at discovering whether you can see how the length of the description reflects the length of the journey. R.S. Thomas wants to show how *remote* the farmhouse is, because one of the things that he is saying about the old woman is that she also is remote. It's a *long* track, and even when you're only part of the way up the sea is '*far*' away. The phrase that really creates this effect, though, is the description of the steep track as 'buttressed with cloud'. This seems to suggest that the track is so high that the clouds are below it – it's almost like the approach to a magical castle in a fairy story.

What are your impressions of [the old woman]?

At the simplest level we know that the woman is old and frail, but the question talks about 'impressions' so it's obvious that something more is called for. She's clearly eager for company, the phrase 'awaits your coming' implies that much. She also wants news, but she's out of touch with how things have changed. The village she knows about is 'in her memory only', she lives in the past. Despite the difficulties presented by the last line it needs to be tackled in your answer to this question. Old age is sometimes associated with wisdom, but R.S. Thomas has turned this idea on its head. The woman is no longer wise, because she lives in a different world. What she has to say is irrelevant to the world the poet has travelled from.

What do you see as the gulf ('abyss') that Thomas feels there is between the old woman and himself? How does he make you feel the width of this gulf?

This question is linked to the last one. The word 'abyss' suggests a gulf that is so deep and wide that there can be no way of crossing it. Indeed, R.S. Thomas mentions that 'no bridge joins her own world with yours'. He creates a sense of how impossible it is to get across with a phrase that shows how hopeless it would be even to try: 'all you can do . . . '. The 'abyss' is not real, of course, it is an image designed to suggest how far apart these two people are. It's partly a matter of her age, partly of where she lives. Both contribute to the feeling that her world is one that no longer exists outside her own imagination.

Your answers to the last two questions are much less closely directed. In **(e)**, though, you would probably have to avoid the two simple alternatives – 'sad? glad?' – that are suggested in the question. The feeling is a bit more complicated than that. After all, the poet himself doesn't tell us all about the old woman just in order to gain our sympathy for her – the last line is fairly brutal. He doesn't have any illusions about what she is like.

If you've been checking this out against the things that you might have said about the poem, you will already have made a start in understanding how to read a poem carefully for an unseen. You'll probably be aware of how different it is from answering a general question on a set book. Instead of turning the text into a bundle of facts and ideas that have to be learnt, what you have to do is to read and re-read until you have thoroughly soaked yourself in the experience that is being described and can imagine it for yourself. Only then can you start to answer questions about what the passage means.

The problem of writing down what you feel is made less daunting by helpful questions like the ones provided in the example. You should still try not to commit yourself to paper until you are ready for it. Make a decision about how much time you can afford to spend reading and stick to it. Some people also find it helpful to doodle all over the text as a way of sorting out their thoughts before writing. Lines can be drawn across the page to show the connections between similar ideas or uses of language, and key words underlined. At the end of such a process, 'Ninetieth Birthday' might look something like this.

You go up the (long) track
That will take a car, but is best walked
On slow foot, noting the lichen
That writes <u>history</u> on the page
Of the grey rock. Trees are about you
At first, but yield to the green bracken,
The nightjar's house: you can hear it spin — *quiet*
hot — On (warm evenings;) it is (still) now
In the (noonday heat,) only the lesser
Voices sound, blue-fly and gnat
And the stream's (whisper.) As the road climbs,
You will pause for breath and the far sea's
Signal will flash, till you turn again
To the (steep) track, (buttressed with cloud.)

remote, as
if in a
fairy tale

And there (at the top) that old woman,
Born almost a century back
In that stone farm, awaits your coming;
Waits for the news of the lost village
She thinks she knows, <u>a place that exists</u>
<u>In her memory only.</u>

                                You bring her greeting
And praise for having lasted so long
With time's knife shaving the bone.
Yet no (bridge) joins her own
World with yours, all you can do
Is lean kindly across (the abyss)
To hear <u>words that were once wise.</u>

R.S. Thomas

It is sometimes said that close reading 'destroys' a poem, and indeed it's difficult to get enthusiastic about a piece of writing when you are sitting in an exam room in a state of suppressed agitation, with a clock ticking away in the corner. You should find, however, that under better circumstances a close reading repays the effort that you put into it. You need time, after all, to sense the full impact of a phrase or image — you can't expect an instant return.

Open book questions

Compare these two questions:

What indications are to be found in the play that the character of Lady Macbeth is not exclusively evil?

and

Re-read Act 1, scene 5. In the evidence of the words of this scene only, does the final description of Lady Macbeth as 'the fiend-like Queen' seem justified? (MEG)

The first question is typical of many exam papers, the second is taken from the MEG specimen question-paper for the GCSE. At first glance, there doesn't seem to be much to choose between them — they're both about Shakespeare's Lady Macbeth and they both ask you to consider how 'evil' or 'devilish' she is. There, however, the resemblance ends. The second question is taken from an 'open book' exam. The play is there in front of you and you can turn to Act 1, scene 5 and re-read it as instructed. That means you have to approach the question in a completely different way. All your prepared notes and quotes on the character of Lady Macbeth take second place to the actual process of reading the scene and trying to understand what motivates her.

You'd probably answer the traditional exam question by starting with your own ideas, and then trying to remember quotes to fit them. When the text is all there for you, the job changes. You have to *start* from the words on the page, and use the essay to explore what they mean. Put another way, this means that instead of wading in with *generalisations* first and looking for particular pieces of evidence later, you start with the detail and only then try to make general points.

Don't fall into the trap of thinking that because you can take your books in with you, the exam will be a push-over. In some ways, open book exams, like coursework, are *more* demanding because you won't get credit for simply having a good memory. A second, closer look at that question on *Macbeth* illustrates the point very well. Here's the beginning of the all-important scene:

SCENE FIVE

Enter LADY MACBETH, *reading a letter*

LADY MACBETH: 'They met me in the day of success; and I have learned by the perfect'st report, they have more in them than mortal knowledge. When I burned in desire to question them further, they made themselves air, into which they vanished.

Whiles I stood rapt in the wonder of it, came missives from the King, who all-hailed me "Thane of Cawdor", by which title, before these weird sisters saluted me, and referred me to the coming on of time with "Hail King that shalt be!" This have I thought good to deliver thee, my dearest partner of greatness, that thou mightst not lose the dues of rejoicing by being ignorant of what greatness is promised thee. Lay it to thy heart, and farewell.'

Glamis thou art, and Cawdor, and shalt be
What thou art promised; yet do I fear thy nature,
It is too full o' th' milk of human kindness
To catch the nearest way. Thou wouldst be great,
Art not without ambition, but without
The illness should attend it. What thou wouldst highly,
That wouldst thou holily; wouldst not play false,
And yet wouldst wrongly win. Thou'dst have, great Glamis,
That which cries 'Thus thou must do, if thou have it';
And that which rather thou dost fear to do
Than wishest should be undone. Hie thee hither,
That I may pour my spirits in thine ear,
And chastise with the valour of my tongue
All that impedes thee from the golden round,
Which fate and metaphysical aid doth seem
To have thee crowned withal.

If you came to that unprepared, you'd quickly become unstuck. Even to understand the scene, you would need:

to know what had happened in the play up to that point;
to understand the language.

To answer the question successfully, it would also be necessary for you:

to know what happens to Lady Macbeth in the rest of the play;
to be able to read well enough to understand not just what she says but why she is saying it.

If you know that Lady Macbeth is reading a letter from her husband who has been told by three witches that he will, sometime in the future, become king of Scotland, you have somewhere to start. You also need to know that Macbeth is a loyal subject of the present king, Duncan, and would appear to have no chance of inheriting the throne in the normal course of events. One possibility that appears already to have occurred to him is assassination.

As you read through the scene, the question that you've been asked should constantly be prompting you into a closer examination of what

Lady Macbeth says and does. It should provide a way of helping you to cross-examine the text yourself. Here are one or two of the things that you might ask as you go through:

1 What does the letter suggest about how Macbeth has reacted to the witches' prediction?
2 What difference might the way in which the news is delivered to Lady Macbeth have on the way she reacts to it?
3 What sort of relationship do Macbeth and Lady Macbeth have?
4 What is Lady Macbeth's immediate reaction to the news?
5 What view of power and greatness does Lady Macbeth have?

The answers that you assemble to questions like these will all help you to re-create the scene in your head, so that the words on the page come alive and you can start to understand Lady Macbeth rather more fully. Only when you've achieved this will you be expected to come to more general conclusions about her.

It's not just in open book exams that you'll come across questions like this. They're becoming increasingly popular with examiners, and lengthy extracts from set books with questions for comment are now likely to be found both on traditional papers and in coursework folders. Naturally, the wording and style of these questions will vary, but the basic aim will remain the same — to give you a chance of working from the text that you are studying rather than a memorised version of it.

SECTION 7

Conclusion

Throughout this book, the emphasis has been on what is new about the GCSE. There's one thing, however, that is unlikely to change, at least in the foreseeable future: the exam will continue to be used by employers in the same way as 'O' level and CSE. There may be a new grading system, but it couldn't, by any stretch of the imagination, be described as startlingly different from the old one. To the surprise of nobody, grades A, B and C at 'O' level have become grades A, B and C at GCSE, whilst CSE grades 2, 3, 4 and 5 have been altered to D, E, F and G. There are no prizes for guessing which grades employers will be looking for, and which grades you will be trying to achieve. The pressure, when it comes, will feel much the same as it always has done.

It's much easier to offer sympathy than help. A lot of the advice that is handed out (and in the summer, just before exams, there's plenty of it) will just leave you feeling inadequate or guilty or both. The one thing you can be sure of is that whatever you are doing, it won't be right.

It's worth finishing, then, not with a list of 'do's and 'don't's about how to revise, but by trying to describe the kinds of candidates who would stand as good a chance as anybody of doing well.

First of all, they'll set standards for themselves. They won't be satisfied with praise if they feel they don't deserve it, nor will they be defeated by criticism and throw in the towel. If they come across something they don't understand, they'll start asking questions and they won't stop until they are positive they understand. They'll be systematic in their approach to their work, sticking to a routine and not departing from it without some very good reason. They won't allow themselves to be intimidated by the sheer scale of what they have to do, but will find a way of reducing it to manageable proportions. Above all, they'll make sure that they know what they want to achieve and what they have to do to achieve it.

If you have found *Countdown to GCSE: English* helpful, it is because you have started to think like this.